SpringerBriefs in Electrical and Computer Engineering

More information about this series at
http://www.springer.com/series/10059

SpringerBriefs in Electrical and Computer Engineering

Mahsa Derakhshani • Tho Le-Ngoc

Cognitive MAC Designs for OSA Networks

 Springer

Mahsa Derakhshani
Department of Electrical
and Computer Engineering
University of Toronto
Toronto, Ontario
Canada

Tho Le-Ngoc
Department of Electrical and Computer Engineering
McGill University
Montreal, Québec
Canada

ISSN 2191-8112   ISSN 2191-8120 (electronic)
SpringerBriefs in Electrical and Computer Engineering
ISBN 978-3-319-12648-7   ISBN 978-3-319-12649-4 (eBook)
DOI 10.1007/978-3-319-12649-4
Springer Cham Heidelberg New York Dordrecht London

Library of Congress Control Number: 2014956040

Printed on acid-free paper

Springer is part of Springer Science+Business Media (www.springer.com)

Preface

The limitations of current static spectrum management policy drive the idea of a more dynamic access policy to improve the efficiency of radio spectrum usage and accommodate the increasing demand for wireless communication applications. Known as the opportunistic spectrum access (OSA), the new paradigm allows cognitive secondary users (SUs) to access the licensed spectrum, provided that the licensed primary users (PUs) are sufficiently protected. To enable OSA by SUs, cognitive medium access control designs are required to track instantaneous spectrum resources, decide on the optimal transmission strategy, and facilitate the distributed spectrum sharing among SUs. This Springer Brief aims to report recent advances in the cognitive MAC designs for OSA networks. First, an overview of the basic MAC functionalities and also MAC enhancements of IEEE 802.11 is presented as an enabler for comprising and realizing full cognitive features. Next, existing MAC protocols for OSA are discussed in detail and classified based on their characteristic features. Finally, new results are presented with regards to adaptive learning-based MAC designs tailored for OSA, which optimize spectrum utilization and ensure a peaceful coexistence of licensed and unlicensed systems. Analytically devised via optimization and game-theoretic approaches, these adaptive MAC designs are shown to effectively reduce collisions between PUs and SUs, and also contention among SUs.

The target audience of this informative and practical Springer Brief is researchers and professionals working on current and next-generation wireless access networks. The content is also valuable for advanced students interested in wireless communications and signal processing for communications.

We would like to acknowledge the financial supports from the Natural Sciences Engineering Research Council of Canada (NSERC) Strategic Projects Grant. Finally, we dedicate this work to our families.

Montréal, Canada
November 2014

Mahsa Derakhshani
Tho Le-Ngoc

Contents

Acronyms

A-BFT	Association beamforming training
AC	Access category
ACK	Acknowledgment
AIFSN	Arbitration inter-frame space number
AT	Announcement time
ATIM	Ad-hoc traffic indication message
A-MPDU	Aggregated MAC protocol data unit
A-MSDU	Aggregated MAC service data unit
AP	Access point
AWGN	Additive white Gaussian noise
BA	Block acknowledgment
BC	Backup channel
BI	Beacon interval
BS	Base station
BSS	Basic service set
BTI	Beacon transmission interval
CAQ	Channel availability query
CBAP	Contention-based access period
C-MAC	Cognitive medium access control
CPE	Customer premises equipment
CSM	Channel schedule management
CSMA	Carrier sense multiple access
CSMA/CA	Carrier sense multiple access with collision avoidance
CTS	Clear-to-send
CVS	Contact verification signal
DCF	Distributed coordination function
DIFS	Distributed inter-frame space
DS	Downstream
DSC	Diagonally strict concavity
EDCA	Enhanced distributed channel access
FCC	Federal Communications Commission

FCH   Frame control header
FCS   Frame control sequence
GDB   Geolocation database
GDD   Geolocation database dependent
HCF   Hybrid coordination function
HCCA   HCF controlled channel access
IEEE   Institute of Electrical and Electronics Engineers
ISM   Industrial, scientific and medical
KKT   Karush-Kuhn-Tucker
LIFS   Long inter-frame space
MAC   Medium access control
MIMO   Multiple input multiple output
M-MAC   Multi-channel medium access control
NCC   Network channel control
NE   Nash equilibrium
OFDM   Orthogonal frequency-division multiplexing
OFDMA   Orthogonal frequency-division multiple access
OSA   Opportunistic spectrum access
OR   Logical OR operation
PBSS   Personal basic service set
PCF   Point coordination function
PCP   PBSS control point
PHY   Physical layer
PMSE   Program making and special event
POMDP   Partially observable Markov decision process
PU   Primary user
QoS   Quality-of-service
RC   Rendezvous channel
RIFS   Reduced inter-frame space
RLQP   Registered location query protocol
RLSS   Registered location secure server
RTS   Request-to-send
SC   Single carrier
SCH   Superframe control header
SIFS   Short inter-frame space
SIP   Spectral image of primary users
SNR   Signal-to-noise ratio
SP   Service period
SRP   Scan result packet
STA   Station
SU   Secondary user
TDM   Time division multiplexing
TDMA   Time division multiple access
TVWS   TV white space
TXOP   Transmit opportunity

UHF	Ultra high frequency
UCS	Urgent coexistence situation
US	Upstream
VHF	Very high frequency
WRAN	Wireless regional area network
WLAN	Wireless local area network
WiMAX	Wireless Interoperability for Microwave Access
WM	Wireless microphone

Chapter 1
Opportunistic Spectrum Access: An Overview

1.1 Static Spectrum Access Limitations

Over the last two decades, there has been a growing interest in wireless communication devices and applications, hence, an ever-increasing demand for radio spectrum. Since radio spectrum is an open and shared broadcast medium, all radios operating over the same frequency band and within the same geographical location interfere with each other. Conventionally, to avoid co-channel interference, the static spectrum allocation strategy is deployed in which different frequency bands are licensed to different types of wireless users who will have an exclusive right to use that portion of the spectrum. However, as the number of wireless users has been exponentially increasing, the spectrum allocation chart has become severely crowded and the availability of frequency bands for emerging wireless applications has become limited.

Furthermore, recent studies [1–4] have indicated that the conventional static spectrum allocation strategy leads to significant spectrum underutilization. More specifically, it has been revealed that while almost all frequency bands have been assigned to licensed users, at any given time and any given location, many frequency slots are unoccupied in a licensed frequency band. In other words, despite the activity of licensed users, measurements show that there still exists plenty of instantaneous spectrum availabilities (also referred to as "spectrum opportunities", "spectrum holes" and "white spaces") in the licensed spectrum. Figure 1.1 illustrates an example of such spectrum holes in both frequency and time domains in a licensed frequency band.

1.2 Cognitive Radio and Opportunistic Spectrum Access

The limitations of static resource allocation on one hand and the success of wireless technologies in unlicensed bands on the other hand demonstrate the need for a more

© The Author(s) 2014
M. Derakhshani, T. Le-Ngoc, *Cognitive MAC Designs for OSA Networks*,
SpringerBriefs in Electrical and Computer Engineering, DOI 10.1007/978-3-319-12649-4_1

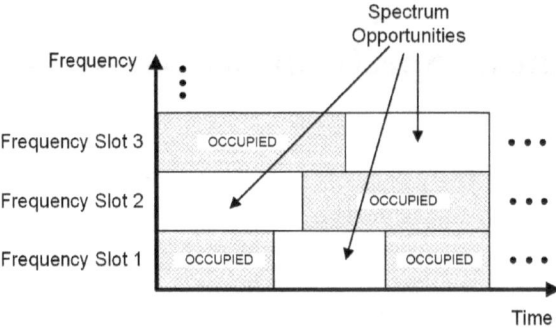

Fig. 1.1 Example of spectrum holes in time and frequency domains in a licensed frequency band

flexible spectrum allocation strategy that could enable further growth for wireless communications. This motivates the idea of opportunistic spectrum access (OSA) to exploit such spectrum availabilities, aiming to improve spectrum utilization [5–9]. OSA allows the unlicensed users (also referred to as secondary users or SUs) to identify and utilize instantaneous spectrum opportunities, while limiting the level of interference to the high-priority licensed users (also referred to as primary users or PUs). Therefore, the two fundamental elements of OSA are opportunity identification and opportunity exploitation. In the opportunity identification (also known as spectrum sensing), secondary users (SUs) need to identify and track dynamically changing idle frequency slots (or channels) in an intelligent way. Based on the observations obtained from the opportunity identification, the opportunity exploitation is responsible to determine the optimal transmission strategy for the secondary access in an idle channel [7].

The key enabling technology for OSA is cognitive radio. More specifically, a cognitive radio is an intelligent and reconfigurable wireless communication system that enables monitoring the radio environment, learning, and accordingly, adapting transmission parameters in order to achieve the optimal spectrum utilization [10–12]. Cognitive radios, together with opportunistic spectrum access, attempt to overcome the dilemma between the increasing spectrum requirements and the scarce spectral resources.

1.3 Emerging Cognitive Radio Standards

In the past few years, there have been worldwide efforts on developing new spectrum policies to accelerate opportunistic usage of spectrum. In particular, in 2004, the Federal Communications Commission (FCC) in the USA allowed unlicensed operations in the unused TV broadcast bands [13], and later released the rules and the technical conditions for such transmission in rural and urban areas using fixed

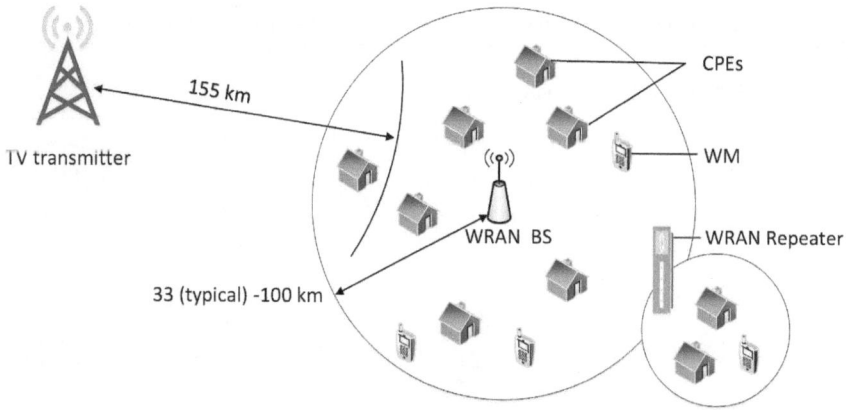

Fig. 1.2 An example of IEEE 802.22 architecture [20]

and portable devices [14, 15]. The white space in the TV band (TVWS) is viewed as one of the first opportunities to adopt and realize the OSA model. This is because of its relatively high robustness to interference (due to the high power transmission) and convenience in opportunity identification (due to the fixed location of TV broadcast sites). Furthermore, the analogue to digital TV switchover results in a reduction of the required spectrum and generates a considerable amount of vacant spectrum in the TV bands [3,4].

In order to take advantage of these spectrum availabilities, new wireless standards are being developed. Some developing standards for using TVWS are namely IEEE 802.22 for wireless regional area networks (WRAN) [16–18] and IEEE 802.11af for wireless local area networks (WLAN) [19–21]. These developing standards are actually extensions of two existing standards, namely IEEE 802.16 Wireless Interoperability for Microwave Access (WiMAX) and IEEE 802.11 WLAN (or as commercially called "Wi-Fi"). They provide incremental improvements on the existing standards towards comprising and realizing full cognitive features [22].

1.3.1 IEEE 802.22 for Wireless Regional Area Networks

IEEE 802.22 WRAN is designed to provide broadband access in rural areas for fixed cognitive radio devices that operate in unused channels in the very-high-frequency/ultra-high-frequency (VHF/UHF) TV bands.

- **Network Architecture:** As shown in Fig. 1.2, IEEE 802.22 WRAN specifies an infrastructured cellular network which includes the base station (BS) and the end-user devices, called customer premises equipment (CPE), or SUs in this case. The coverage range of each BS is specified between 30 and 100 km. In order to

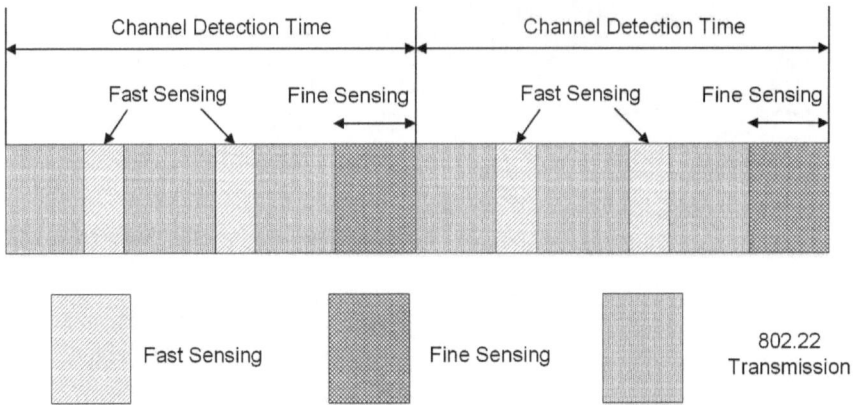

Fig. 1.3 Two-stage sensing mechanism in IEEE 802.22 [23]

extend the coverage and network capacity, the repeaters can be deployed in the 802.22 WRANs [17,20]. Figure 1.2 illustrates a sample architecture for the IEEE 802.22 WRAN system.

- **Cell Association:** To join the network, a CPE first should search for an operating WRAN cell. Then, it needs to sense the operating channels of the corresponding BS, which are declared in the superframe control header (SCH). Upon a free channel detection, the CPE will inform the BS, acquire the location information, and synchronize itself to the BS by using the superframe preamble [18].
- **Incumbent Protection:** Currently, the TVWS band is licensed to TV broadcast/multicast services and also broadcast auxiliary services such as program making and special event (PMSE) users. Typical examples of PMSE users are low-power devices to support news broadcasting and theatrical productions, for example wireless microphones, whose operation is allowed by regulatory authorities. To protect PUs in TVWS band, both geolocation databases (GDBs) and periodic spectrum sensing are considered to be used in order to enable a PU-SU coexistence. The GDB is an authorized database that stores and keeps track of the allowable TVWS frequency channels and their corresponding operating parameters for any geographic location [21]. The BS periodically receives updates from GDB and ensures that the associated CPEs are not interfering with PUs.

In addition, to fully protect PUs, the local periodic spectrum sensing is also performed by CPEs. During spectrum sensing, all CPEs operating on the same frequency channel should stop transmission. The 802.22 MAC can manage periodic spectrum sensing by informing CPEs about quiet periods in SCH. The CPEs report their sensing measurements to the BS during urgent coexistence situation (UCS) intervals in upstream subframes. Through cooperative sensing, BS would notify the CPEs about the PU returns and switch the frequency channel to avoid any interference [18].

As shown in Fig. 1.3, in IEEE 802.22, periodic spectrum sensing includes two stages of fast sensing and fine sensing. Fast sensing is performed every 1 ms

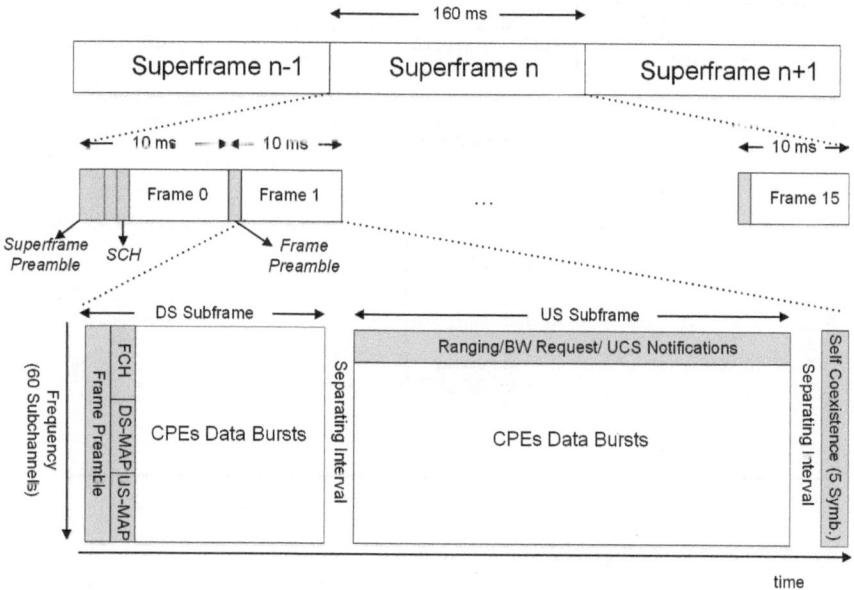

Fig. 1.4 Frame and superframe structure in IEEE 802.22 [18, 23]

in each channel. Nevertheless, depending on the results of fast sensing, a more accurate sensing result might be required. Consequently, fine sensing will be performed to improve the detection performance. Fine sensing exploits a longer sensing duration than fast sensing, and uses feature detection techniques to increase the sensing accuracy [23].

- **MAC Layer:** In the IEEE 802.22 standard, each BS is responsible for the resource allocation among the CPEs within its cell. The scheduling-based MAC mechanism of IEEE 802.22 is a combination of time division multiple access (TDMA) and orthogonal frequency division multiple access (OFDMA), with the frame hierarchy as illustrated in Fig. 1.4 [17, 18]. In 802.22, both upstream (US) data transmissions from CPEs to BS and downstream (DS) data transmissions from BS to CPEs are scheduled within MAC *frame* units with equal length of 10 ms. Each frame includes two *subframes* for US and DS data transmissions. Multiple MAC frames (up to 16), each preceded by the frame preamble who is responsible for synchronization and channel estimation, are concatenated to build a *superframe* [18]. Each superframe begins with a superframe preamble, which is used by CPEs for synchronization, and also the SCH, which is used by BS in order to publicize its current available channels, its supported bandwidths, and quiet periods among CPEs [23].

To allocate the available resources in time and frequency domain among CPEs, BS schedules packet bursts to/from different CPEs within the DS and US subframes. The scheduling information are distributed by the BS in DS/US-MAPs in the DS subframe. Furthermore, at the start of DS subframe, there is a frame

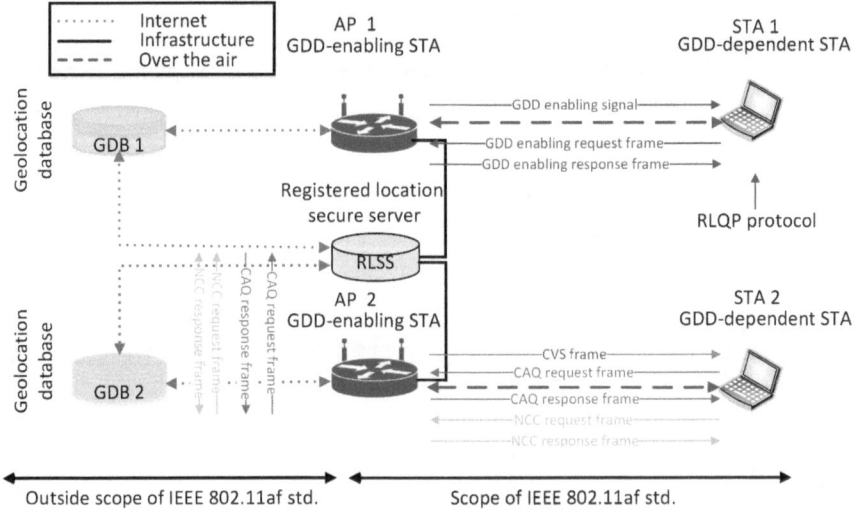

Fig. 1.5 An example of IEEE 802.11af architecture [21, 27]

control header (FCH), which announces the size of the DS-MAP and US-MAP fields and also includes channel descriptors. US subframes include UCS in which BS notifies CPEs if any PU has been detected in the operating channel. In addition, ranging (i.e., deriving the CPE distance from the BS) and bandwidth requests are performed in the US subframes [18, 23].

Although IEEE 802.22 is the first international cognitive radio standard, considering the deployment cost of WRAN infrastructure, it still remains unclear whether the WRAN could create a profitable service in rural areas [16–25].

1.3.2 IEEE 802.11af for Wireless Local Area Networks

Furthermore, as another potential application for OSA, IEEE 802.11af (also known as "White-Fi" or "Wi-Fi 2.0") is being developed to enhance the capacity and quality-of-service (QoS) of current Wi-Fi systems with the use of the TVWS. Due to the better propagation characteristics of the VHF/UHF bands as compared to the industrial, scientific and medical (ISM) band, it is expected that IEEE 802.11af will offer higher speed and wider coverage than the current Wi-Fi, and also will support better QoS guarantees [20].

- **Network Architecture:** In the current IEEE 802.11 WLAN architecture, the basic building block (also known as a basic service set or BSS) consists of an access point (AP) and its associated stations (STAs). However, IEEE 802.11af is a wireless network comprising four different structural entities including GDB, registered location secure server (RLSS), and two geolocation database dependent (GDD) stations, which are GDD-enabling and GDD-dependent STAs. As discussed in the previous section, GDB is an authorized database, which contains recent information of available TVWS bands and corresponding operating parameters at any given location and time. RLSS is a local database, which includes the geographic location and current channel use information for all BSSs that communicate with it. Furthermore, RLSS can provide faster responses on the permitted operating parameters to the BSSs under its control. The GDD STAs are those whose operations depend on information received from an authorized GDB [21, 26, 27].

Two types of GDD stations are introduced in 802.11af, i.e., GDD-enabling and GDD-dependent STAs. A GDD-enabling STA is in fact an AP with extended functionalities, which can securely communicate with GDBs in addition to controlling its associated GDD-dependent STAs. Based on the acquired information from GDB, a GDD-enabling STA generates and transmits a white space map (WSM) which represents a list of available TVWS and corresponding power limitations for GDD-dependent STAs under its operating region. Thus, a GDD-dependent STA, or the SU in this case, represents a STA in the BSS architecture, which is under the control of a GDD-enabling STA [21, 26, 27]. Figure 1.5 illustrates an example of 802.11af architecture with two BSSs.

- **Incumbent Protection:** In 802.11af, incumbent protection is performed through communication with regulated GDBs. Given the limited capabilities of SUs (i.e., STAs) in spectrum sensing, collecting information from an authorized entity such as GDB provides a more secure approach [21]. However, the local spectrum sensing can be considered as a complementary technique to ensure a reliable SU-PU coexistence.

- **MAC Layer:** 802.11af mainly defines the communication protocol between the GDD-dependent STAs, GDD-enabling STAs and RLSS. This protocol is called registered location query protocol (RLQP), which is provided to share WSMs and channel utilization among GDD STAs. Based on this communication, STAs can choose operating parameters such as frequency channel, power, and bandwidth. It should be noted that RLSS and GDD-enabling STAs communicate with GDB through internet, and their communication procedure is outside the scope of 802.11af standard [21].

A GDD-enabling STA transmits a beacon frame, also called *GDD-enabling signal*, in-band on a chosen available TVWS channel to inform that it offers GDD enablement service, and hence, to form a BSS. After receiving this beacon signal, a GDD-dependent STA enables and responds by sending *GDD enablement response* frame [21, 27]. In each BSS, 802.11af standard adopts a CSMA-based channel access scheme for spectrum sharing between different GDD-dependent STAs in a BSS.

Several other MAC mechanisms are defined in 802.11af, such as channel availability query (CAQ), channel schedule management (CSM), contact verification signal (CVS), and network channel control (NCC). Such mechanisms help to share and keep updated WSMs among GDD-dependent STAs, and manage channel utilization among GDD-enabling STAs in a neighborhood. More specifically, the GDD STAs, either GDD-dependent or GDD-enabling STAs, use CAQ procedure to query RLSS or other GDD-enabling STAs to obtain WSMs. CSM requests are transmitted only by a GDD-enabling STA to a RLSS or other GDD-enabling STAs to update the channel schedule information of other GDD-enabling STAs [21, 27].

The CVS is sent by a GDD-enabling STA aiming to ensure that its associated GDD-dependent STAs are within its reception range, and also validate their operation under a correct WSM. NCC is a two-way handshake procedure to manage frequency channel utilization. First, a NCC request is transmitted by a GDD STA to announce its preferred frequency channels permitted by WSM. Then, through a NCC response, RLSS or the GDD-enabling STA grants the permission for usage if available, and indicates the transmit power constraints [21, 26, 27].

According to [22], modern WLANs could already be considered as cognitive radios because of their coexistence capabilities and dynamically changing frequencies and transmission power. The basic coexistence capability of IEEE 802.11 is its listen-before-talk MAC based on the carrier sense multiple access with collision avoidance (CSMA/CA). The simplicity and success of the listen-before-talk contention-based MAC has the potential to be applied for the early cognitive systems [22].

1.4 Challenges in MAC Design for OSA Networks

Opportunistic spectrum access causes a revolutionary change in the radio spectrum regulations and also plays an important role in enhancing the spectrum usage efficiency. That is because it enables time-varying and flexible usage of radio spectrum, while taking into account technical and regulatory considerations. The spectrum opportunities are time-varying depending on the activities of PUs in the licensed channels (e.g., TVWS are dynamic due to digital TV multicasting and activity of low-power devices such as wireless microphones). This fact makes an OSA network a highly dynamic and challenging wireless environment. To get the most out of such dynamic spectrum opportunities while protecting the spectrum licensees from interference, it is critical to design a cognitive MAC protocol that is capable of filling the spectral gaps of PUs in an intelligent way. Thus, the need for intelligent coexistence capabilities and autonomous coordination in OSA raises a new set of technical challenges which are not present in the existing radio systems. In the following, we discuss main technical challenges for cognitive MAC design in OSA networks.

1.4.1 Adapting to Primary User Transmission

Modeling the dynamic behavior of the PUs and the interactive behavior of SUs in response to PUs' dynamics is a major design issue in an OSA network. To provide sufficient benefit to SUs, an accurate model is needed to simultaneously capture the dynamics of spectrum opportunities and describe the decision process of SUs. Furthermore, since cognitive radio can be viewed as an enabler for distributed radio resource management, it is important for such a model to be able to support distributed operations of SUs, in which each SU autonomously coordinates its usage and independently achieves the optimal transmission strategy.

1.4.2 Compatibility with Primary Users

To design a cognitive MAC scheme for SUs, there is a tradeoff between two conflicting objectives: maximizing the opportunistic spectrum utilization for SUs and minimizing the possible collisions between SUs and PUs. To identify the time-varying spectrum holes for transmission, SUs need to periodically monitor the channels to determine whether or not PUs are active. Although SUs are intended to only exploit the spectrum holes for communications, two types of collision between SUs and PUs could happen. First, a SU may miss-detect the PU's activity and start transmitting while the PU is present. Such a miss-detection is due to inevitable errors in spectrum sensing which are caused by noise and fading. Second, even if an idle channel is perfectly detected and used by a SU, a collision may still occur since the PU may return and reoccupy that idle channel at any time during the SU transmission. Figure 1.6 illustrates these interference situations between a SU and a PU in a given frequency slot. In the presence of miss-detection errors and non-zero PU return probabilities, in order to reduce the effects of collisions between PUs and SUs, an adaptive transmission strategy needs to be adopted for SUs. As such, SUs need to optimally configure their transmission parameters in order to simultaneously minimize collisions and make the best use of the available opportunities.

To guarantee the compatibility with legacy systems, it is important to define and impose an appropriate interference constraint to sufficiently protect PUs' communications. This is because different definitions of the interference constraint may result in different levels of protection for PUs. Generally, an interference constraint must reflect two parameters: first, the maximum tolerable interference level at an active primary receiver and second, the maximum tolerable probability that the interference level exceeds its maximum level [7]. However, to precisely define an interference constraint, there are other aspects that need to be addressed. For example, in an OSA network with multiple SUs, the locations of SUs with respect to the primary receiver and the channel propagation characteristics must be considered to set a constraint on the aggregate interference caused by SUs to a PU [28].

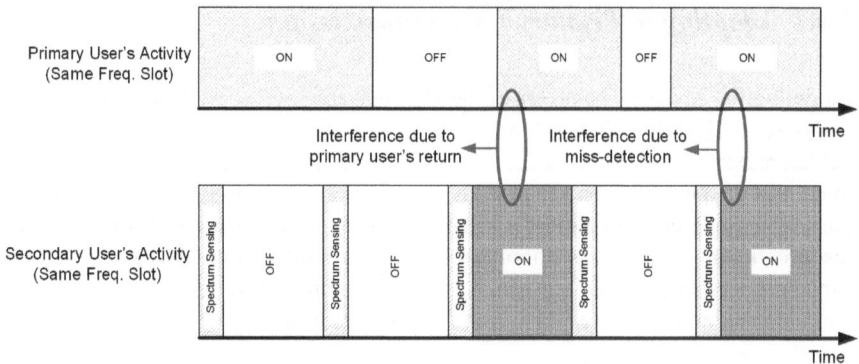

Fig. 1.6 Interference situations between a SU and a PU

1.4.3 Information Exchange Between Secondary Users

In the multi-user MAC design in OSA networks, a key challenge is how to coordinate and share opportunities among SUs to achieve a network-level objective. To efficiently manage spectrum sharing and also achieve an accurate spectrum sensing, each SU may need to collect and exchange information about the spectrum usage pattern of PUs and other SUs. However, such information exchange constitutes an overhead in either infrastructureless or infrastructure-based networks. Thus, reliable control channels need to be devised. Alternatively, for the distributed MAC design, learning-based algorithms can be developed in which system parameters are adapted to their optimal values over time based on the locally available information.

1.5 Structure of This Brief

The objective of this Springer-Brief is to present OSA networks and explore recent advances in cognitive MAC designs in such a network. The remainder of this brief is organized into the following chapters.

Chapter 2 presents an overview on the MAC mechanisms and enhancements currently deployed in IEEE 802.11 standards as enablers for full cognitive MAC designs. Then, we will review relevant studies on the MAC design in OSA networks that will be used for the development of cognitive MAC schemes in the subsequent chapters.

Chapter 3 studies and develops cognitive MAC protocols in OSA networks for SUs, taking into account the time-varying and dynamic behavior of PUs. Aiming to reduce the effects of collision between PUs and SUs due to the PU random return, a transmission strategy is proposed for SUs to dynamically hop over multiple idle frequency slots, each with an adaptive activity factor to be determined.

Taking into account the spectrum sharing among SUs, the dynamic PU activity and channel characteristics, the SU activity factor optimization problem for maximizing the overall SU throughput is formulated. More specifically, two sets of constraints are considered to support a fair access and enable orthogonal spectrum sharing in the time domain among individual SUs. Furthermore, the throughput of each SU is defined as the successful transmission rate. Subsequently, the optimal MAC algorithm is developed based on the dual decomposition method. By introducing an adaptive carrier sense multiple access (CSMA) scheme, a learning-based distributed access algorithm for SUs is proposed. Then, using the stochastic gradient descent optimization framework, an analysis is provided on the convergence properties of the proposed algorithm.

Chapter 4 is concerned with the game-theoretic design of the activity factor optimization problem, while using the proposed adaptive CSMA scheme for spectrum sharing among SUs. Via the game theory framework, it is shown that the formulated game is an exact potential game. The conditions assuring the feasibility and optimality of the Nash equilibrium (NE) in this non-cooperative design are examined. To achieve the globally optimum solution, learning algorithms including best-response dynamics and log-linear dynamics are developed. In addition, to reflect the inherent competition among SUs in the adaptive CSMA scheme and to enable contention control, a different design objective is introduced for the activity factor optimization problem. Since the new problem is non-convex, obtaining its globally optimal solution is highly intractable. Thus, the MAC design is cast into a game-theoretic framework that enables a distributed implementation and fast convergence. In the formulated game, each SU selfishly determines its optimal activity factors, without any coordination with other SUs. In this competitive design, it is shown that the NE may be highly inefficient compared to the globally optimal solution of the fully coordinated design. To improve the efficiency of such an NE, the SUs are forced to act in a more cooperative manner by applying a dynamic pricing mechanism. Finally, to reach the steady state, an iterative algorithm based on best-response dynamics is proposed.

Chapter 5 evaluates the performance of the proposed adaptive CSMA scheme by developing an analytical model to compute the system throughput in a single channel. The system throughput is defined and derived as the fraction of opportunities used to successfully transmit data, taking into account the inevitable collisions among SUs. It is shown that the adaptive CSMA scheme is able to intelligently control the contention among the STAs, while effectively reducing the collision probability. Thus, it achieves a higher throughput compared to the conventional CSMA. This is because the adaptive CSMA allocates a higher chance of transmission to those STAs with more favorable conditions through granting higher activity factors. Moreover, the effects of network configuration parameters (e.g., the number of STAs in the network, the minimum contention window, and the packet length) on the throughput performance are investigated.

References

1. FCC Spectrum Policy Task Force, "Report of the spectrum efficiency working group," Federal Communications Commission, Tech. Rep. ET Docket No. 02–135, Nov. 2002.
2. M. H. Islam, C. L. Koh, S. W. Oh, X. Qing, Y. Y. Lai, C. Wang, Y.-C. Liang, B. E. Toh, F. Chin, G. L. Tan, and W. Toh, "Spectrum survey in Singapore: Occupancy measurements and analyses," in *Proc. IEEE Intl. Conf. on Cognitive Radio Oriented Wireless Networks and Commun. (CROWNCOM)*, Singapore, May 2008.
3. K. Harrison, S. M. Mishra, and A. Sahai, "How much white-space capacity is there?" in *Proc. IEEE Intl. Symp. New Frontiers in Dynamic Spectrum Access Networks (DySPAN)*, Singapore, Apr. 2010.
4. J. V. D. Beek, J. Riihijarvi, A. Achtzehn, and P. Mahonen, "TV white space in Europe," *IEEE Trans. Mobile Comput.*, vol. 11, no. 2, pp. 178–188, Feb. 2012.
5. J. Mitola, "Cognitive radio for flexible mobile multimedia communications," in *Proc. IEEE Intl. Workshop on Mobile Multimedia Commun. (MOMUC)*, San Diego, CA, USA, Nov. 1999.
6. I. F. Akyildiz, W. Y. Lee, M. C. Vuran, and S. Mohanty, "NeXt generation/dynamic spectrum access/cognitive radio wireless networks: A survey," *Computer Networks (Elsevier)*, vol. 50, no. 13, pp. 2127–2159, Sep. 2006.
7. Q. Zhao and B. M. Sadler, "A survey of dynamic spectrum access: Processing, networking, and regulatory policy," *IEEE Signal Process. Mag.*, vol. 24, no. 3, pp. 79–89, May 2007.
8. I. F. Akyildiz, W. Y. Lee, M. C. Vuran, and S. Mohanty, "A survey on spectrum management in cognitive radio networks," *IEEE Commun. Mag.*, vol. 46, no. 4, pp. 40–48, Apr. 2008.
9. B. Jabbari, R. Pickholtz, and M. Norton, "Dynamic spectrum access and management," *IEEE Wireless Commun. Mag.*, vol. 17, no. 4, pp. 6–15, Aug. 2010.
10. J. Mitola, *Software Radios: Wireless architecture for the 21st century.*   New York: Wiley, 2000.
11. S. Haykin, "Cognitive radio: Brain-empowered wireless communications," *IEEE J. Sel. Areas Commun.*, vol. 23, no. 2, pp. 201–220, Feb. 2005.
12. A. Goldsmith, S. A. Jafar, I. Maric, and S. Srinivasa, "Breaking spectrum gridlock with cognitive radios: An information theoretic perspective," *Proc. IEEE*, vol. 97, no. 5, pp. 894–914, May 2009.
13. FCC, "Notice of proposed rulemaking," ET Docket No. 04–113, May 2004.
14. FCC, "Second report and order and memorandum opinion and order," ET Docket No. 08–260, Nov. 2008.
15. FCC, "Unlicensed operations in the TV broadcast bands," Second Memorandum Opinion and Order, ET Docket No. 10–174, Sep. 2010.
16. IEEE Std. 802.22, "Information Technology–Telecommunications and Information Exchange between Systems–Wireless Regional Area Networks (WRAN)–Specific Requirements–Part 22: Cognitive Wireless RAN Medium Access Control (MAC) and Physical Layer (PHY) Specifications: Policies and Procedures for Operation in the TV Bands," pp. 1–680, http://www.ieee802.org/22/, Jul. 2011.
17. C. Stevenson, G. Chouinard, L. Zhongding, H. Wendong, S. J. Shellhammer, and W. Caldwell., "IEEE 802.22: The first cognitive radio wireless regional area network standard," *IEEE Commun. Mag.*, vol. 47, no. 1, pp. 130–138, Jan. 2009.
18. N. Tadayon, and S. Aissa, "Modeling and analysis of cognitive radio based IEEE 802.22 wireless regional area networks." *IEEE Trans. Wireless Commun.*, vol. 12, no. 9, pp. 4363–4375, Sep. 2013.
19. R. Kennedy and P. Ecclesine, "IEEE P802.11af tutorial," IEEE 802.11-10/0742r0, https://mentor.ieee.org/802.11/dcn/10/11-10-0742-00-0000-p802-11af-tutorial.ppt, Jul. 2010.
20. K. G. Shin, H. Kim, A. W. Min, and A. Kumar, "Cognitive radios for dynamic spectrum access: From concept to reality," *IEEE Trans. Wireless Commun.*, vol. 17, no. 6, pp. 64–74, Dec. 2010.
21. A. B. Flores, R. E. Guerra, E. W. Knightly, P. Ecclesine, and S. Pandey, "IEEE 802.11af: A standard for TV white space spectrum sharing." *IEEE Commun. Mag.*, vol. 51, no. 10, pp. 92–100, Oct. 2013.

22. L. Berlemann and S. Mangold, *Cognitive radio and dynamic spectrum access*. Wiley, 2009.
23. C. Cormio and K. R. Chowdhury, "A survey on MAC protocols for cognitive radio networks," *Ad Hoc Networks*, vol. 7, no. 7, pp. 1315–1329, Sep. 2009.
24. C. Ghosh, S. Roy, and D. Cavalcanti, "Coexistence challenges for heterogeneous cognitive wireless networks in TV white spaces," *IEEE Wireless Commun. Mag.*, vol. 18, no. 4, pp. 22–31, Aug. 2011.
25. C. Cordeiro, K. Challapali, D. Birru, and N. Sai Shankar "IEEE 802.22: The first world-wide wireless standard based on cognitive radios," in *Proc. IEEE Intl. Symp. New Frontiers in Dynamic Spectrum Access Networks (DySPAN)*, Baltimore, MD, USA, Nov. 2005.
26. K. H. Chang"IEEE 802 standards for TV white space," *IEEE Wireless Commun.*, vol. 21, no. 2, pp. 4–5, Apr. 2014.
27. D. Lim"TVWS regulation and standardization (IEEE 802.11af)", ACT LAB. LG Electronics, May 2013.
28. M. Derakhshani and T. Le-Ngoc, "Aggregate interference and capacity-outage analysis in a cognitive radio network,", *IEEE Trans. Veh. Technol.*, vol. 61, no. 1, pp. 196207, Jan. 2012.

Chapter 2
Cognitive MAC Designs: Background

This chapter first presents an overview on the MAC mechanisms currently deployed in IEEE 802.11 WLANs. The basic coexistence capabilities and recent enhancements of 802.11 MAC are discussed as enablers for realizing full cognitive MAC designs. Then, the second part of this chapter reviews various state-of-the-art cognitive MAC designs in OSA networks, which serve as the background for the development of cognitive MAC designs in subsequent chapters. We discuss and categorize the MAC design approaches in OSA networks, considering the need for network-wide coordination, the network structure of SUs, and the transmission model of PUs.

2.1 IEEE 802.11 MAC Protocol as Enabler

The IEEE 802.11-based WLANs are becoming more popular and widely deployed around the world. One of the main reasons for such success is the robust and flexible MAC protocol with coexistence capabilities. In IEEE 802.11 standard, the basic MAC mechanism has two different operation modes: distributed coordination function (DCF) and *optional* point coordination function (PCF). PCF is a centralized MAC protocol in which a centralized scheduler at the AP coordinates access among different STAs by sending polling messages, aiming to support collision-free services. However, DCF is a contention-based access scheme, based on CSMA/CA using binary exponential backoff rules to manage retransmission of collided packets [1]. The uncoordinated yet reliable access mechanism of DCF made it the fundamental MAC mechanism of 802.11. In the following, the operation of listen-before-talk DCF is briefly reviewed, as an initial step toward more intelligent spectrum access schemes.

© The Author(s) 2014   15
M. Derakhshani, T. Le-Ngoc, *Cognitive MAC Designs for OSA Networks*,
SpringerBriefs in Electrical and Computer Engineering, DOI 10.1007/978-3-319-12649-4_2

2.1.1 Distributed Coordination Function

DCF requires a STA, with a new packet for transmission, to sense the channel activity prior to transmission. If the channel is sensed idle for a time interval equal to a distributed inter-frame space (DIFS), the STA transmits. Otherwise, if the STA senses a transmission either immediately or during the DIFS, it continues monitoring the channel. When the channel is measured idle for a DIFS, the STA backoffs for a random period of time. The backoff mechanism enables collision avoidance by minimizing the probability of collision with other STAs. Furthermore, a STA must go through the backoff mechanism between two consecutive packet transmissions to avoid the channel capture [1].

DCF uses a discrete-time backoff mechanism, i.e., the time following a DIFS is slotted. The backoff time-slot length needs to be designed equal to the time a STA requires to detect the transmission of a packet from any other STA. At each packet transmission, the backoff time is selected according to a uniform distribution in the interval $(0, W - 1)$ where W represents the contention window which is a function of the number of transmissions already failed for the packet. Each STA starts the packet transmission by setting W equal to the minimum contention window size (i.e., CW_{min}). According to the binary exponential backoff rules, W is doubled after each unsuccessful transmission. Each STA increases W up to the maximum contention window size $CW_{max} = 2^m CW_{min}$ where m represents the maximum backoff stage [1].

The backoff time counter is decremented and a STA transmits when the backoff time counter reaches zero. Once the data packet is received successfully, the receiver waits for a period of time called short inter-frame space (SIFS), and then sends an acknowledgment (ACK). By sensing the ACK, the receiver informs the transmitter about the successful reception of the transmitted packet. If the ACK is not received by the transmitter, it retransmits that packet according to the exponential backoff rules [1].

To improve the throughput performance of CSMA/CA in IEEE 802.11, in addition to the basic access mechanism, an optional four-way handshaking technique, i.e., request-to-send/clear-to-send (RTS/CTS), has been proposed for a packet transmission. In the RTS/CTS access mechanism, a STA who is ready to transmit, after waiting for a DIFS and passing the backoff process, has to transmit a special short frame called request-to-send (RTS) before transmitting its packet. After detection of the RTS frame by the receiver, it responds by transmitting a clear-to-send (CTS) frame after a SIFS. If the CTS frame is correctly detected by the transmitter, it is allowed to transmit its packet afterwards. The RTS/CTS access mechanism effectively reduces the average collision time because collisions can be early detected by the transmitters when the CTS is not received [1].

Figure 2.1 illustrates an example of the channel-access procedure of two STAs using CSMA/CA with the RTS/CTS access mechanism. At the end of the packet transmission of STA 1, both STAs wait for a DIFS and pick their backoff times. Since the backoff time of STA 2 is shorter, it wins the competition and starts the packet transmission, while STA 1 is still in the middle of its backoff procedure.

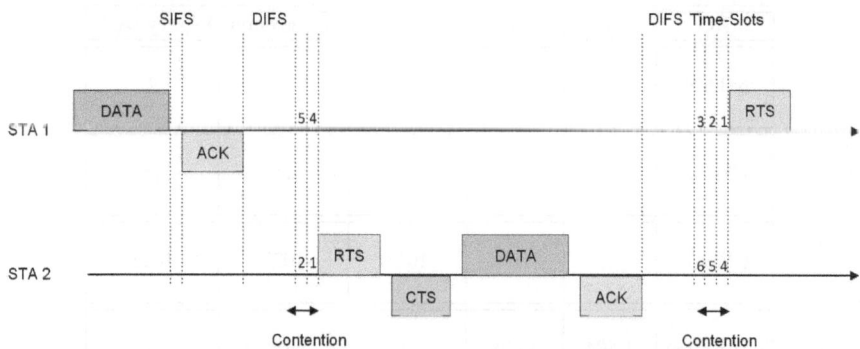

Fig. 2.1 Example of the channel-access procedure of two STAs using CSMA with the RTS/CTS access mechanism in time domain

When STA 1 senses the channel busy because of the transmitted RTS, it stops its backoff mechanism. When the channel is measured idle again for a DIFS, STA 1 joins the competition and sets its backoff time to 3 without resetting its backoff counter. However, STA 2 randomly picks a new backoff time (i.e., 6).

Although the binary exponential backoff mechanism is effective in controlling collision among STAs, the coexistence capabilities of DCF are limited. For instance, taking into account QoS support, DCF fails to adequately meet the performance requirements of voice and video applications, since it was initially developed only for best effort services. As a result, in high traffic scenarios, the delay increases significantly for different types of traffic using DCF as traffic load goes high. Thus, in the following section, we review MAC enhancements introduced by different 802.11 amendments and standards, to improve the overall throughput of the network and provide QoS guarantee for real-time multimedia applications.

2.1.2 MAC Enhancements

2.1.2.1 IEEE 802.11e for Quality-of-Service Support

IEEE 802.11e is an amendment to the IEEE 802.11 base standard which introduces significant QoS support features. To provide better QoS provisioning, it defines hybrid coordination function (HCF), which is an enhanced MAC protocol by introducing two different access mechanisms, i.e., HCF controlled channel access (HCCA) and enhanced distributed channel access (EDCA). HCF is called hybrid since two proposed access mechanisms, i.e., HCCA and EDCA, cover both centralized contention-free and distributed contention-based control, respectively [2, 3].

HCCA is an improved version of PCF to provide centralized medium access scheduling. Until now, no known device exists that uses HCCA. More popular MAC mechanism is EDCA. EDCA provides traffic differentiation between four different

AC	Voice	Video	Best Effort	Background	Legacy DCF
AIFSN	2	2	3	7	2
CW_{min}	3	7	15	15	15
CW_{max}	7	15	1023	1023	1023
TXOP(ms)	1.504	3.008	0	0	0

Table 2.1 EDCA default parameters for different ACs

access categories (ACs) or traffic classes, which are voice, video, background and best effort. These four types of traffic can have service differentiation based on the following parameters [2, 3].

- **Arbitration inter-frame space number (AIFSN):** By assigning variable waiting times (before transmission) to different ACs, EDCA can prioritize one AC over the other. More specifically, AIFSN represents the required number of backoff time-slots that a STA should wait before either starting a transmission or going through the backoff process [4]. In fact, AIFSN is a variable alternative for fixed DIFS in DCF.
- **Contention window:** Another method for service differentiation is to allow different maximum and minimum contention window sizes to different ACs. Higher priority ACs are assigned smaller CW_{min} and CW_{max}, to ensure smaller backoff times, more frequent transmissions, and hence less delay.
- **Transmit opportunity (TXOP):** To decrease the collision avoidance overhead, frame bursting is offered in 802.11e in which the STA that obtains transmission opportunity (after winning in the backoff competition) can send a burst of back-to-back packets based on its channel quality for a fixed period of time. A STA cannot transmit longer than a TXOP. Consequently, in 802.11e, TXOP length can be varied for different ACs to achieve different levels of priority [5, 6].

Table 2.1 shows the default values of AIFSN, CW_{min}, CW_{max}, and TXOP for four different ACs recommended in the 802.11e draft standard [4].

In addition to the basic QoS functionalities provided by EDCA in 802.11e, there have been several algorithms proposed in the literature to further improve QoS in this standard. In [7], various such techniques have been presented including bandwidth allocation, data control, and distributed admission control to protect on-going high-priority traffic.

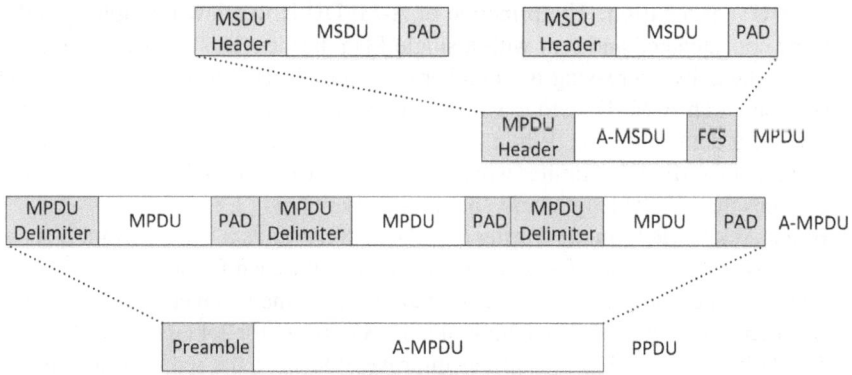

Fig. 2.2 Two-level frame aggregation with A-MSDU and A-MPDU [3]

2.1.2.2 IEEE 802.11n for High Throughput

IEEE 802.11n was released in 2009 in order to improve throughput over earlier standards (i.e., 802.11a and 802.11g). There are major PHY enhancements introduced in 802.11n such as the use of 5 GHz band in addition to the 2.4 GHz, orthogonal frequency division multiplexing (OFDM), and multiple input multiple output (MIMO). Adding 5 GHz, it creates an opportunity to have more channels with 20 MHz as well as 40 MHz bandwidth.

With a large number of packets to be transmitted, signaling load on the channel from PHY and MAC headers can be significant. Furthermore, multiple single packets will require multiple random backoff periods which will decrease the throughput as well. One way to reduce such overhead and also minimize the wasted time in waiting is frame aggregating. In the frame aggregation, by concatenating or packing multiple packets together, overheads can be added over a group of packets rather than over separate ones. Furthermore, there is no need to pass the backoff procedure for every single packet. More specifically, the three major MAC enhancements developed in 802.11n to reduce overheads are aggregated MAC service data unit (A-MSDU), aggregated MAC protocol data unit (A-MPDU), and block acknowledgments (BA), which are explained in the following [3, 8–10].

- **MSDU Aggregation:** The principle of A-MSDU is to concatenate multiple MSDUs destined for the same receiver into a single MPDU. This will require only one MPDU header for all aggregated MSDUs within an A-MSDU. In order to achieve this aggregation, the incoming packets from the link layer are first buffered to collect a number of them before aggregation. Although MSDU aggregation reduces overheads, it will increase delay for these packets which is a drawback for aggregation. As shown in Fig. 2.2, each MSDU has its own headers of source and destination addresses, data packet length, and some padding bits in the end. A-MSDU aggregates only those packets which have the same source and destination addresses [3, 8–10].

- **MPDU Aggregation:** The principle of A-MPDU is to allow multiple MPDUs to be concatenated and sent with a single PHY header. This achieves reduction in overhead by decreasing the number of required PHY headers. Each MPDU consists of an A-MSDU along with a MPDU header and frame check sequence (FCS) for data validation. While forming an A-MPDU, padding bytes are added along with a MPDU delimiter which is added for each MPDU. MPDU delimiter contains the MPDU length field and the signature field [3, 8–10].
- **Block Acknowledgment:** During a TXOP, a STA sends a burst of frames separated by SIFS. Instead of sending back an ACK for each frame, a BA is sent in 802.11n. The transmitter sends a BA request, then the receiver responds with a BA after a SIFS period. The length of a SIFS is 10 μs for 2.4 GHz and 16 μs for 5 GHz band in 802.11n. To further improve efficiency, the reduced inter-frame space (RIFS) has been introduced which is only 2 μs of length, and thus much shorter than SIFS. This results in reduced overheads of frequent ACKs and waiting times [3, 8–10].

2.1.2.3 IEEE 802.11ac for very High Throughput

Although the recent WLAN standard 802.11n can support up to 600 Mbps for a single STA in a BSS, the network-wide throughput is still restricted by the maximum link data rate. To improve the network throughput, IEEE 802.11ac has been developing since 2011 and was approved in January 2014. The major PHY enhancements in this standard are supporting wider channel bandwidths, higher modulation schemes, larger number of MIMO spatial streams, and, last but not least, downlink multi-user MIMO. Enabling channel bonding techniques, in 802.11ac, separate 40 MHz bands can be combined to form one 80 MHz or even 160 MHz channel, which leads to increase the overall throughput to more than 1Gbps. Modulation up to 256 QAM is possible in this standard adding two bits per symbol as compared to 64 QAM in the earlier standard to increase the throughput. Additionally, 8×8 MIMO and downlink multi-user MIMO are supported in order to enable sending multiple packets to multiple STAs simultaneously [3, 11].

The MAC enhancements in this standard mainly deal with multi-user MIMO, wider channel bandwidths, and co-existence with legacy WLANs. One of the most important MAC enhancements is TXOP sharing. TXOP sharing allows multiple downlink traffic streams to be sent to multiple receivers in the same TXOP. There are serious limitations with the legacy EDCA TXOP where frames belonging to a single AC can be sent to a STA in one TXOP. Thus, neither packets belonging to different STAs nor packets belonging to different ACs could not be put in one TXOP. With downlink multi-user MIMO capability, in 802.11ac, TXOP sharing makes it possible to send packets to different ACs of the same STA as well as to different STAs within the same TXOP [3, 12].

Enabling TXOP sharing, in a downlink scenario with different AC queues at AP, each AC uses its own EDCA parameters to compete for a TXOP. Once an AC gets the TXOP, this AC is called the primary AC. After this phase, the primary

AC decides whether secondary ACs are permitted on the same TXOP or not. The primary AC can have multiple STAs as well but only one AC can be the primary AC. The length of this TXOP is also determined and limited by the transmission of the primary AC even though secondary ACs have more packets to send. After the TXOP period is over, each STA sends a BA separately in separate times to ensure packet delivery [3, 12].

In 802.11n, frame aggregation is first introduced by presenting A-MSDU and A-MPDU. To further increase MAC efficiency and improve PHY data rates, in 802.11ac, an enhanced aggregation scheme is proposed, in which maximum size of A-MSDU and A-MPDU are increased [3, 13]. Furthermore, in order to improve coexistence capabilities which is harder with wider channels because of increased overlaps, enhanced secondary channel *clear channel assessment* has been introduced along with a new operating mode notification frame as more explained in [14]. The idea behind this mechanism is that if two STAs are partially interfering with each other while their respective different APs are not aware of it, the interfered STA can notify its respective AP to use only the interference free part of the channel instead of the complete channel. Thus, based on the interference from other systems, a STA might request an AP to reduce the used channels to a subset of the original channel bandwidth where interference is minimal.

2.1.2.4 IEEE 802.11ad for Multi-Gigabit Throughput

802.11ad is an amendment to 802.11 operating in 60 GHz, aiming to support multi-Gigabit wireless communication. This standard has been developing to serve throughput intensive and short-range applications such as multimedia wireless display and local data/file transfer. The motivations to use the 60 GHz band are providing the opportunity to have larger available unlicensed band compared to 2.4/5 GHz, and hence, the chance to use wider channels, reaching up to 7 Gbps transmission data rate [3, 14–16].

Operating at 60 GHz suffers from higher propagation and atmosphere loss compared to 2.4/5 GHz. To compensate the signal attenuation, beamforming can provide a solution allowing the transmitted power to be focused. The small wavelength of 60 GHz facilitates a feasible and efficient implementation of beamforming deploying phased-array antennas. This is because large antenna arrays can be integrated in mobile devices due to their small sizes. In addition to the wider channel, beamforming also helps in increasing throughput while reducing the interference between STAs [14, 15].

The major PHY enhancement provided by 802.11ad is to present a single carrier (SC) modulation and coding scheme in addition to OFDM PHY. This SC PHY is designed to reduce processing power and enable lower complexity transceivers, using shorter symbol structure and simpler coding. Introducing both SC and OFDM modulations provides the flexibility such that OFDM PHY enables high data-rate transmissions up to 7 Gbps in frequency-selective environment, while SC PHY supports over 4.6 Gbps data-rates with low-complexity transceiver [3, 14].

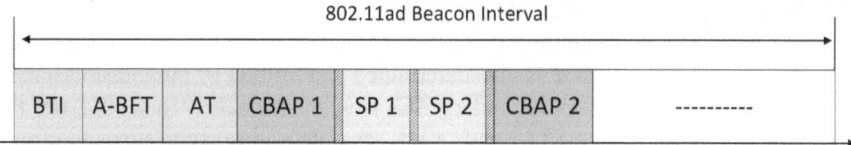

Fig. 2.3 The structure of a beacon interval in 802.11ad [3, 14]

In 802.11ad, MAC layer has been divided into two versions: basic MAC and enhanced MAC. Basic MAC is the same as the legacy 802.11e and 802.11n networks to support STAs from these networks. Enhanced MAC on the other hand is to support beamforming and high throughput through directional communication, including a *hybrid* random and scheduling-based access scheme. To provide a network more adaptable to directional transmission, 802.11ad defines a new structural building block, called personal BSS, which is also more appropriate for its target applications. In the PBSS, one STA takes the role of PBSS control point (PCP) which is responsible to transmit the beacon frames. If no beacons are received by a STA, it may become a PCP and start sending beacons. Similarly at another location, any other STA can act as the PCP. This makes this structure very flexible and adaptable [3, 14].

In a beacon interval (BI), the first phase is beacon transmission interval (BTI), in which a PCP would potentially discover new STAs by sending out beacons in different directions. In the second phase of BI, association beamforming training (A-BFT) is performed between the STA and the PCP to further tune the beamforming between PCP and STA. In the third phase, announcement time (AT), PCP transfers control and management information to all STAs. The last phase is the data transfer time (DTT), including the contention-based access periods (CBAPs) and service periods (SPs) [3, 14]. A structure of a beacon interval is illustrated in Fig. 2.3.

During a CBAP, any STA can access the channel based on the modified 802.11e EDCA with directional medium access rules, where CSMA/CA is used for channel access and aggregation for data as well as acknowledgments. A new aggregation scheme, called video aggregation MSDU (VA-MSDU), has been introduced able to support video traffic. [3, 14].

The SPs are dedicated and scheduled for particular STAs using TDMA. Since certain applications such as wireless display or VoIP have very strict requirements on jitter and delay, in order to fulfill such requirements TDMA has been introduced for SP periods of the beacon frame. Therefore SP uses TDMA for streaming and real time applications sensitive to jitter and delay while CBAP based on CSMA/CA is used for bursty traffic such as internet browsing. TDMA allocates some time-slots in the SP fields to some STAs where they wake up on these fixed time intervals to transmit and receive information. This is also beneficial since TDMA can be used for directional communication using beamforming while CSMA/CA can only operate in omni-directional transmission and reception mode [3, 14].

Standard		802.11e	802.11n	802.11ac	802.11ad
Release		2005	2009	2014	Under Development
Data Rate			600Mbps	1Gbps	7Gbps
PHY Enhancements	Frequency		2.4 GHz 5 GHz	5 GHz	2.4/5 GHz 60 GHz
	Channel BW		20 MHz 40 MHz	20/40 MHz 80 MHz 160 MHz	2160 MHz
	Modulation		B/QPSK 16 QAM 64 QAM	B/QPSK 16/64 QAM 256 QAM	OFDM & SC
	MIMO Streams		4 × 4	8 × 8, Downlink MU-MIMO	-
MAC Enhancements	Service Differentiation	4 Access Categories	-	TXOP Sharing	Beamforming
	Frame Aggregation	Packet Bursting (TXOP)	A-MSDU A-MPDU Block ACK	Larger A-MSDU A-MPDU	VA-MSDU
	Access Mechanism	EDCA	EDCA	EDCA	Directional CSMA+TDMA

Table 2.2 Evolution of PHY and MAC enhancements in recent IEEE 802.11 generations

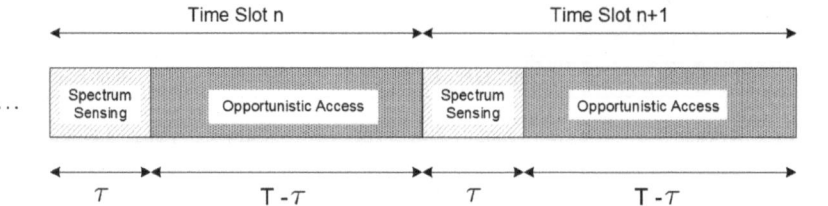

Fig. 2.4 Example of a general 2-phase time-slot structure for SUs

2.1.3 Summary of MAC Enhancements

Table 2.2 summarizes and compares MAC and PHY specifications, and also enhancements in all recent and upcoming 802.11 WLAN standards. The MAC enhancements are categorized into three categories considering proposed service differentiation, frame aggregation, and medium access techniques.

2.2 MAC Protocols for Opportunistic Spectrum Access

Although the MAC mechanisms of 802.11 standards can be already considered as cognitive MAC protocols, OSA requires more intelligent and flexible MAC protocol designs, specifically in the face of QoS support. MAC design in OSA networks includes two key functions, i.e., *spectrum sensing* to identify instantaneous spectrum opportunities, and *spectrum access* to coordinate SUs and protect PUs.

Spectrum sensing is one of the key elements in the establishment of cognitive radio, since its accuracy and response time directly affect the efficiency of opportunistic spectrum access. SUs need to periodically sense the channels to detect spectrum holes and avoid collisions. More specifically, each SU needs to follow a slotted transmission scheme. Each time-slot with an equal duration T consists of two periods: *sensing* of duration τ and *transmission* of duration $(T - \tau)$. Figure 2.4 depicts an example of the general time-slot structure employed by SUs.

Different approaches have been proposed for spectrum sensing in OSA networks, such as matched filtering, energy detection (e.g., in [17]), and feature cyclostationary detection (e.g., in [18]). We refer the interested reader to [19,20] and references therein for reviewing recent advances in spectrum sensing techniques. Though MAC layer is not responsible for adopting a sensing approach, MAC protocols mainly support scheduling of spectrum sensing, e.g., optimizing sensing and transmission time tradeoff.

In the cognitive MAC design in OSA networks, *spectrum access* is responsible to maximize the spectrum utilization of SUs by properly designing their spectrum access strategies, while limiting the conflicts between SUs and PUs. More specifically, spectrum access specifies that which channels are assigned to who, when, and for how long. Different cognitive MAC protocols, including coordinated time-slotted and distributed random-access-based schemes, have been proposed for both centralized and ad-hoc OSA networks in the literature [21, 22]. Thus, to further explore such studies, we categorize the existing MAC protocols in OSA networks into (i) coordination-based protocols, and (ii) random-access-based protocols, considering the need for network-wide coordination and the network structure of SUs. Additionally, we take into account the transmission model of PUs as another design feature in the presented cognitive MAC protocols classification as compared to the surveys in [21, 22].

2.2.1 Coordination-Based Protocols

The coordination-based MAC protocols need to maintain time synchronization throughout the network and perform a large amount of information exchange, aiming to manage resource negotiation and share sensing results.

This protocols are mainly suitable for point-to-multipoint systems, where there exists a central controller, such as a BS, that is responsible for managing the spectrum access of all SUs [25, 26]. More specifically, a central controller synchronizes and coordinates the spectrum allocation and sharing among SUs, aiming to maximize the overall network performance. To obtain the optimal solution, the central controller needs to communicate with SUs in order to know their transmission requirements. Furthermore, it needs to gather and process the information about channel availabilities.

For instance, IEEE 802.22, the BS manages the spectrum allocation and sharing among CPEs within its own cell. In IEEE 802.22 MAC, time division multiplexing (TDM) and demand-assigned TDMA are respectively used in the downstream and upstream directions [21]. More details of MAC mechanism of IEEE 802.22 were discussed in Sect. 1.3.1.

In addition, for ad-hoc networks without the presence of central entities, coordination-based MAC protocols have also been investigated in some recent works (e.g., in [23, 24]), aiming to improve the overall link throughput and increase robustness to frequency channel change. However, to develop a coordination-based MAC protocol in ad hoc OSA networks, there are several design challenges such as providing network-level synchronization and enabling information exchange with minimum overhead possible [21].

For instance, in [23], a rendezvous channel (RC) is defined and used for the proposed coordination-based MAC protocol (also known as *cognitive-MAC* or C-MAC) to enable time synchronization and information sharing among SUs. This RC is actually a control channel and is chosen as the channel which is available for the longest time in the network. SUs periodically visit the RC to get resynchronized, announce any channel change, and obtain the recent information on PU and other SUs activities in each channel. Each channel is structured into superframes, each composed of two phases, i.e., the beacon period and the data transmission period.

In the C-MAC protocol, the quiet and beacon periods for each of the frequency channels are non-overlapping using RC, thus each SU would get a chance to sense any channel or transmit a beacon on any channel. By processing the SU traffic and reservation information collected from beacons, load balancing is performed for each superframe. The major limitation of C-MAC is that it suffers from unscalability and control channel starvation, since all the beacons transmitted by SUs must be included in the beacon period of RC. It is also not sufficiently clear how C-MAC can coordinate quiet and beacon periods in different channels to stay non-overlapping, specifically for a larger number of channels. Furthermore, energy efficiency is not taken into consideration in C-MAC [21, 23].

Aiming to achieve the energy efficient communication, another coordination-based MAC protocol is proposed for ad-hoc OSA networks in [24], by allowing SUs to enter a low-power doze state with no communication. Furthermore, the proposed *multi-channel MAC* (M-MAC) protocol exploits an energy-efficient two-stage spectrum sensing mechanism with a low-power inaccurate sensing (or fast sensing) and a high-power accurate sensing (or fine sensing). To perform network synchronization and share control information, M-MAC uses a dedicated out-of-band control channel.

In the M-MAC protocol, time is divided into two periods, including the ad-hoc traffic indication message (ATIM) window and the DATA window. During the ATIM window, each SU participates in synchronization process, performs fast sensing, shares sensing results, and performs a two-way handshake for medium access coordination. During the DATA window, each SU transmits data and also performs fine sensing if required [22, 24].

When a SU connects to the OSA network, it performs fast sensing over all the channels during a ATIM window, and stores the sensing results in a vector called spectral image of PUs (SIP). If a SIP value of a channel shows uncertainty, the SU will schedule a fine sensing in the following DATA window. Afterwards, in each ATIM, each SU randomly selects one of the channels to update its SIP. Subsequently, SUs share their sensing outcomes and cooperate to detect spectrum opportunities through a subframe in the ATIM window. This subframe includes C time-slots, each representing one channel. SUs transmit a busy tone in a time-slot if their SIP does not show inactive PU in the corresponding channel. Applying an OR fusion rule, if a time-slot is detected busy, the corresponding channel will be consequently excluded for data exchange during the DATA window. The subframe synchronization is managed by a scan result packet (SRP) which is transmitted on the control channel by a SU. SRP indicates the beginning of the cooperative sensing subframe in the ATIM window [22, 24].

In order to access the channel, a SU with buffered packets transmits a ATIM packet on the control channel, which contains the information of the preferred channel and its queue status. Then, upon agreement on the selected channel, the receiver responds back by sending an ATIM ACK. SUs who exchanged ATIM packets, stay awake and communicate during DATA window. The SUs not involved in any data exchange enter a doze state to save power until next ATIM window. Although the

M-MAC protocol is effective to be energy efficient, its reliability is significantly sensitive to control channel quality and synchronization precision [22, 24].

In summary, the difficulties in keeping an ad-hoc network synchronized and designing a reliable common control channel make it almost infeasible to adopt a coordination-based MAC protocol in ad-hoc OSA networks. Thus, more *distributed* designs are required, in which the spectrum access decisions need to be made independently by each SU. Corresponding to different design objectives, the way SUs manage spectrum allocation could be cooperative or non-cooperative. In a cooperative design, each SU optimizes its spectrum access strategy, aiming to enhance the overall performance of the network. However, in a non-cooperative design, each SU maximizes its own benefit, without being concerned about the overall network performance [25, 26]. In the distributed OSA MAC design literature, both ways are explored. On one hand, optimization and cooperative game-theoretic approaches are devised, enabling SUs to achieve a network-level objective (e.g., [27]). On the other hand, non-cooperative game-theoretic designs are applied in which each SU selfishly maximizes its own performance (e.g., [28, 29]).

In distributed OSA MAC schemes, a common way of sharing the spectrum by SUs is to use random access schemes. In the following section, we review the proposed random-access-based cognitive MAC protocols in OSA networks, which can be classified considering their assumptions on the transmission model of PUs.

2.2.2 Random-Access-Based Protocols

A random-access-based MAC protocol should be able to perform the channel contention and reservation to manage the spectrum sharing among SUs. Random access-based MAC protocols are developed in multi-channel and multi-user OSA networks [30–37]. In the following, we review existing random-access-based MAC protocols in OSA networks, considering the transmission model of PUs.

In the cognitive MAC design literature, two different transmission structures, i.e., time slotted and un-slotted transmission, are considered for PUs. To develop MAC protocols for SUs, several studies assume that both PUs and SUs have the same transmission time-slot structure [30, 38–41]. In this case, the collision between SUs and PUs occurs only due to the miss-detection errors. With such assumption of synchronous slotted transmission structure, time coordination is required between PU and SU networks. Nevertheless, since time coordination between PU and SU networks cannot be feasible in many scenarios, the transmission of PUs is assumed to be un-slotted in [42–45]. Accordingly, the traffic pattern of PUs is modeled as a continuous-time ON/OFF random process with the ON (1) and OFF (0) states respectively representing the *busy* and *idle* periods of the PU. In this case, even if an idle channel is perfectly detected and used by a SU, a collision can still happen since the PU may return and reoccupy that idle channel at any time during the SU transmission.

2.2.2.1 Time Slotted Transmission Structure for Primary Users

Assuming a synchronous slotted transmission structure between PU and SU networks, in [30], an optimal design of OSA is developed. To address hardware limitations and energy costs, it is considered that the SU cannot sense all channels at a certain time. Consider a frequency band licensed to PUs, which is divided into N_p non-overlapping frequency slots (or channels), each with bandwidth $B^i, i = 1, \ldots, N_p$. Also, assume a secondary network with N_s SUs looking for temporal spectrum opportunities in these N_p channels.

In particular, it is assumed that each SU can sense and access only Q_1 channels in each time-slot, where $Q_1 < N_p$. Prior to start spectrum sensing at the beginning of each time-slot, a SU needs to select a subset $\mathscr{A}_s$ ($|\mathscr{A}_s| \leq Q_1$) of channels to sense. Then, given the sensing observations for channels in $\mathscr{A}_s$, the SU picks a subset $\mathscr{A}_a$ ($\mathscr{A}_a \subset \mathscr{A}_s$) of the sensed channels to access [30].

In [30], it is assumed that the PU activities in different channels follow a discrete-time Markov process with $M = 2^{N_p}$ states, while each state represents ON/OFF status on all N_p channels.

Since the SU is able to sense only parts of the available channels, the overall state of the network is partially observable. Thus, the joint design of sensing and access strategies of a SU is modeled as a partially observable Markov decision process (POMDP) [46]. In the formulated POMDP, the actions of a SU are sensing and access strategies (i.e., $\{\mathscr{A}_s, \mathscr{A}_a\}$) and the observations of a SU are the results of its sensing. According to the decision and observation history, a SU updates a belief vector which reflects its knowledge about the system state. More specifically, the belief vector of a SU is represented by $\Omega(t) = [\omega_1(t), \ldots, \omega_M(t)]$ where $\omega_i(t)$ is the probability that the network is in the state i at the beginning of time-slot t.

In [30], when a SU senses some channels ($\mathscr{A}_s$) and transmits in a subset of them ($\mathscr{A}_a$), it then receives a reward which is defined as the number of transmitted bits, i.e.,

$$r(t) = \sum_{i \in \mathscr{A}_a} S_i(t) B^i \tag{2.1}$$

where $S_i(t) \in \{0,1\}$ is the state of channel i in time-slot t. Subsequently, with a constraint on the maximum collision probability between a SU and a PU, an optimization problem is designed to maximize the expected total number of transmitted bits in N_{ts} time-slots, i.e.,

$$\max \mathrm{E}\Big[\sum_{t=1}^{N_{ts}} r(t) | \Omega(1) \Big]$$

$$\text{subject to } P_{col} \leq \zeta \tag{2.2}$$

where P_{col} is the probability of collision between the SU and the PU, ζ is the the maximum tolerable probability of collision by the PU, and $\Omega(1)$ is the initial belief vector.

The optimal solution of this POMDP model can be obtained using a linear programming algorithm. However, as the number of channels (i.e., N_p) increases, the computational complexity grows exponentially, due to the state space growth. Thus, by reducing the number of states, a suboptimal strategy is also devised with reduced complexity [30].

In [30], a simple mechanism is considered for the random sharing among SUs based on the CSMA scheme with an in-band signaling. Nevertheless, the effect of contention among SUs is not reflected in the optimal design of access strategies for SUs. In the proposed method, each SU picks actions independently and competes randomly to capture the channel.

In [31, 32], SUs perform contention on an out-of-band control channel prior to the optimization process. In each time-slot, the winner of the contention continues to find its optimal sensing and access strategy. Since the wining SU is able to sense only a subset of channels (i.e., $\mathscr{A}_s$), there might be spectrum opportunities that are overlooked and wasted in each time-slot [47].

In addition to the sensing and access strategies, the design of the spectrum sensor operating characteristics is studied in [38], under the assumption of erroneous spectrum sensing. In [30, 38], it is assumed that the channel occupancy follows a Markovian model for which the state transition probabilities are known to the SUs. However, such information may not be available for the SUs.

Assuming that the statistics of the channel availabilities are not available a priori, the studies in [39–41] investigate learning-based OSA approaches for SUs that have partial sensing ability. In [39,40], the design of optimal sensing and access strategies is formulated as a multi-armed bandit process [48], where there is a tradeoff between using well-explored channels and searching unexplored channels. The study in [41] extends the earlier work in [39] by considering sensing errors.

2.2.2.2 Un-Slotted Transmission Structure for Primary Users

Assuming no time coordination between SUs and PUs, in [42], a transmission strategy for a single SU is proposed to prevent a possible collision between the SU and the PU due to the random return of the PU. In the proposed access scheme for the SU, the transmission duration within each time-slot (i.e., $T - \tau$) is divided into S sub-slots. The SU can transmit data in the first n_t consecutive sub-slots. Then, it does not transmit in the remaining $S - n_t$ sub-slots (see Fig. 2.5a). This transmission strategy helps to reduce the collision probability between the SU and the PU. This is because the PU return probability during the SU transmission is an increasing function of time, given that channel is detected idle in the sensing duration. Subsequently, n_t is optimized based on the PU traffic models, while bounding the collision probability between the SU and the PU below a target level.

In [43], the OSA design is studied for a single SU that shares spectrum with a data-centric PU network (e.g., GSM networks and 802.11-based WLANs). To take advantage of the short-lived opportunities created between the packet bursts in such PU network, a different transmission strategy is proposed for the SU. In the proposed access scheme, the SU can transmit in each separate sub-slot during a

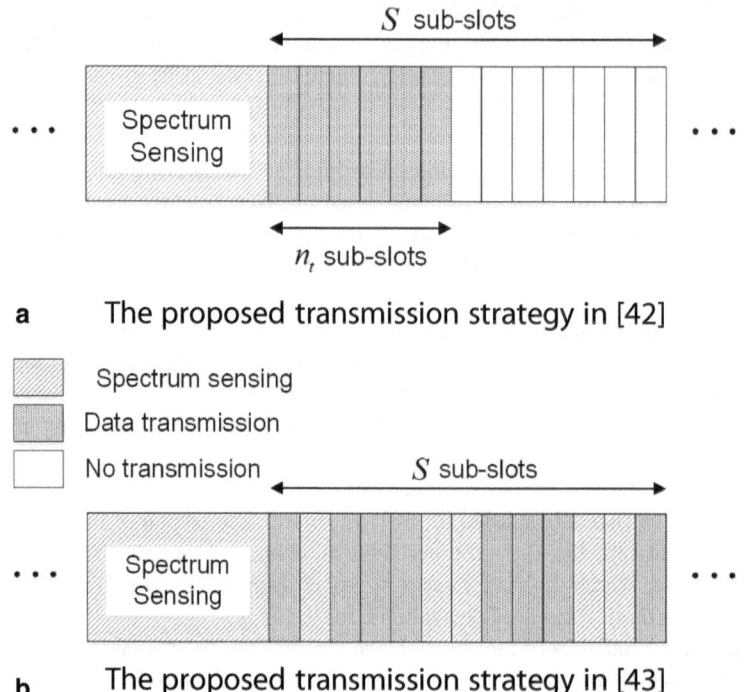

a   The proposed transmission strategy in [42]

b   The proposed transmission strategy in [43]

Fig. 2.5 Example of transmission strategies for a single SU proposed in **a** [42], **b** [43]

transmission duration (see Fig. 2.5b). Compared to [42], the proposed access strategy in [43] allows more flexibility for the SU to optimize its spectral efficiency. Using a POMDP framework, the SU decides to either access or perform extra spectrum sensing in each sub-slot, by maximizing its probability of successful transmission over S sub-slots.

For a multi-channel scenario, [44] presents an OSA strategy for a single SU. In the proposed method, the SU periodically senses the channels, assuming that the SU can sense only one channel in each sensing phase. Based on the spectrum sensing history, the SU decides either to transmit on one of the channels or not to transmit in each transmission phase. By learning from the sensing outcomes in different channels, the optimal channel to access is selected by maximizing the SU throughput, while limiting the collision probability caused by a SU to the PU due to sensing errors and PU random return.

When studying cognitive MAC design under the effect of PU return, past OSA MAC studies mainly focus on designing optimal channel access schemes, considering a single SU or a simple mechanism for random sharing among SUs without taking into account the competition or coordination among SUs. Thus, further research is needed to develop OSA MAC schemes that apply some sharing incentives among

SUs to avoid the channel degradation due to the crowding effects. This coordination can lead to a higher network performance compared to when each SU picks actions independently and competes randomly to capture the channel.

2.3 Concluding Remarks

In this chapter, we have provided a survey on the available techniques in the cognitive MAC design for SUs in OSA networks. It has been discussed that the assumption about the transmission structure of PUs plays an important role. Only a limited number of works, such as [42–45], have considered the OSA MAC design under the assumption of no time coordination between SUs and PUs. However, these proposed schemes mainly focus on the design for a single SU. Furthermore, in an ad-hoc OSA network, there is a need for a fully distributed random-access based MAC design with the cooperative behavior for SUs. These observations motivate us to study the distributed random-access-based MAC design and present the performance analysis in a multi-user OSA network, considering the effect of random returns of PUs and contention among SUs. Recent advances in cognitive MAC designs for an OSA network in [49–52] will be covered intensively in this Springer Brief.

References

1. G. Bianchi, "Performance analysis of the IEEE 802.11 distributed coordination function," *IEEE J. Sel. Areas Commun.*, vol. 18, no. 3, pp. 535–547, Mar. 2000.
2. Y. Xiao, "IEEE 802.11e: A QoS provisioning at the MAC layer," *IEEE Wireless Commun.*, vol. 11, no. 3, pp. 72–79, Jun. 2004.
3. E. Charfi, L. Chaari, and L. Kamoun, "PHY/MAC enhancements and QoS mechanisms for very high throughput WLANs: A survey," *IEEE Commun. Surveys Tuts.*, vol. 15, no. 4, pp. 1714–1735, Fourth Quarter 2013.
4. IEEE Std 802.11e, "Wireless LAN medium access control (MAC) and physical layer (PHY) specifications: Amendment 8: Medium Access Control (MAC) Quality of Service enhancements," *IEEE Computer Society*, 2005.
5. I. Tinnirello and S. Choi, "Efficiency analysis of burst transmissions with block ACK in contention-based 802.11e WLANs," in *Proc. IEEE Intl. Conf. Commun. (ICC)*, Seoul, Korea, May 2005.
6. P. K. Hazra and A. De, "Performance analysis of IEEE 802.11e EDCA with QoS enhancements through TXOP based frame-concatenation and block-acknowledgement," *Intl. J. Adv. Tech.*, vol. 2, no. 4, pp. 542–560, 2011.
7. Y. Xiao, "QoS guarantee and provisioning at the contention-based wireless MAC layer in the IEEE 802.11e wireless LANs," *IEEE Wireless Commun.*, vol. 13, no. 1, pp. 14–21, Feb. 2006.
8. D. Skordoulis, Q. Ni, H. Chen, A. Stephens, C. Liu, and A. Jamalipour, "IEEE 802.11n MAC frame aggregation mechanisms for next-generation high-throughput WLANs," *IEEE Wireless Commun. Mag.*, vol. 15, no. 1, pp. 40–47, Feb. 2008.
9. Y. Lin and V. W. Wong, "WSN01-1: Frame aggregation and optimal frame size adaptation for IEEE 802.11n WLANs," in *Proc. IEEE Global Commun. Conf. (GLOBECOM)*, San Francisco, CA, USA, Nov. 2006.

10. Y. Kim, S. Choi, K. Jang, and H. Hwang, "Throughput enhancement of IEEE 802.11 WLAN via frame aggregation," in *Proc. IEEE Veh. Tech. Conf. (VTC)*, Los Angeles, CA, USA, Sep. 2004.

11. O. Bejarano, E. W. Knightly, and M. Park, "IEEE 802.11 ac: From channelization to multi-user MIMO," *IEEE Commun. Mag.*, vol. 51, no. 10, pp. 84–90, Oct. 2013.

12. C. Zhu, Y. Kim, O. Aboul-Magd, C. Ngo, "Multi-user support in next generation wireless LAN," in *Proc. Consumer Commun. Netw. Conf. (CCNC)*, Las Vegas, NV, USA, Jan. 2011.

13. E. H. Ong, J. Kneckt, O. Alanen, Z. Chang, T. Huovinen, and T. Nihtil, "IEEE 802.11ac: Enhancements for very high throughput WLANs," in *Proc. Symp. on Personal, Indoor and Mobile Radio Commun. (PIMRC)*, Toronto, ON, Canada, Sep. 2011.

14. E. Perahia, M. X. Gong, "Gigabit wireless LANs: an overview of IEEE 802.11ac and 802.11ad," *ACM SIGMOBILE Mobile Comput. Commun. Review*, vol. 15, no. 3, pp. 23–33, Jul. 2011.

15. L. Verma, M. Fakharzadeh, and S. Choi, "Wifi on steroids: 802.11ac and 802.11ad," *IEEE Wireless Commun.*, vol. 20, no. 6, pp. 30–35, Dec. 2013.

16. E. Perahia, C. Cordeiro, M. Park, and L. L. Yang, "IEEE 802.11 ad: Defining the next generation multi-Gbps Wi-Fi," in *Proc. Consum. Commun. Netw. Conf. (CCNC)*, Las Vegas, NV, USA, Jan. 2010.

17. F. F. Digham, M. S. Alouini, M. K. Simon, "On the energy detection of unknown signals over fading channels," *IEEE Trans. Commun.*, vol. 55, no. 1, pp. 21–24, Jan. 2007.

18. M. Derakhshani, T. Le-Ngoc, M. Nasiri-Kenari, "Efficient cooperative cyclostationary spectrum sensing in cognitive radios at low SNR regimes," *IEEE Trans. Wireless Commun.*, vol. 10, no. 11, pp. 3754–3764, Nov. 2011.

19. T. Yucek and H. Arslan, "A survey of spectrum sensing algorithms for cognitive radio applications," *IEEE Commun. Surveys Tuts.*, vol. 11, no. 1, pp. 116–130, First Quarter 2009.

20. E. Axell, G. Leus, E. Larsson, and H. Poor, "Spectrum sensing for cognitive radio: State-of-the-art and recent advances," *IEEE Signal Process. Mag.*, vol. 29, no. 3, pp. 101–116, Apr. 2012.

21. C. Cormio and K. R. Chowdhury, "A survey on MAC protocols for cognitive radio networks," *Ad Hoc Networks*, vol. 7, no. 7, pp. 1315–1329, Sep. 2009.

22. A. De Domenico, E. C. Strinati, and M. Di Benedetto, "A survey on MAC strategies for cognitive radio networks," *IEEE Commun. Surveys Tuts.*, vol. 14, no. 1, pp. 21–44, First Quarter 2012.

23. C. Cordeiro and K. Challapali, "C-MAC: A cognitive MAC protocol for multichannel wireless networks," in *Proc. IEEE Intl. Symp. New Frontiers in Dynamic Spectrum Access Networks (DySPAN)*, Dublin, Ireland, Apr. 2007.

24. M. Timmers, S. Pollin, A. Dejonghe, L. Van der Perre, and F. Catthoor, "A distributed multichannel MAC protocol for multihop cognitive radio networks," *IEEE Trans. Veh. Technol.*, vol. 59, no. 1, pp. 446–459, Jan. 2010.

25. I. F. Akyildiz, W. Y. Lee, M. C. Vuran, and S. Mohanty, "A survey on spectrum management in cognitive radio networks," *IEEE Commun. Mag.*, vol. 46, no. 4, pp. 40–48, Apr. 2008.

26. E. Hossain, D. Niyato, and Z. Han, *Dynamic Spectrum Access and Management in Cognitive Radio Networks*.   Cambridge, 2009.

27. K. Liu and Q. Zhao, "Cooperative game in dynamic spectrum access with unknown model and imperfect sensing," *IEEE Trans. Wireless Commun.*, vol. 11, no. 4, pp. 1596–1604, 2012.

28. D. Niyato and E. Hossain, "Competitive spectrum sharing in cognitive radio networks: A dynamic game approach," *IEEE Trans. Wireless Commun.*, vol. 7, no. 7, pp. 2651–2660, Jul. 2008.

29. M. Maskery, V. Krishnamurthy, and Q. Zhao, "Decentralized dynamic spectrum access for cognitive radios: Cooperative design of a non-cooperative game," *IEEE Trans. Wireless Commun.*, vol. 57, no. 2, pp. 459–469, Feb. 2009.

30. Q. Zhao, L. Tong, A. Swami, and Y. Chen, "Decentralized cognitive MAC for opportunistic spectrum access in ad hoc networks: A POMDP framework,"*IEEE J. Sel. Areas Commun.*, vol. 25, no. 3, pp. 589–600, Apr. 2007.

31. H. Su and X. Zhang, "Cross-layer based opportunistic MAC protocols for QoS provisionings over cognitive radio wireless networks," *IEEE J. Sel. Areas Commun.*, vol. 26, no. 1, pp. 118–129, Jan. 2008.
32. J. Jia, Q. Zhang, and X. Shen, "HC-MAC: A hardware-constrained cognitive MAC for efficient spectrum management," *IEEE J. Sel. Areas Commun.*, vol. 26, no. 1, pp. 106–117, Jan 2008.
33. A. C. C. Hsu, D. S. L. Wei, and C.-C. J. Kuo, "A cognitive MAC protocol using statistical channel allocation for wireless ad-hoc networks," in *Proc. IEEE Wireless Commun. Netw. Conf. (WCNC)*, Kowloon, Hong Kong, Mar. 2007.
34. H. Su and X. Zhang, "Channel-hopping based single transceiver MAC for cognitive radio networks," in *Proc. IEEE Conf. on Inf. Sciences and Systems (CISS)*, Princeton, NJ, USA, Mar. 2008.
35. B. Hamdaoui and K. G. Shin, "OS-MAC: An efficient MAC protocol for spectrum-agile wireless networks," *IEEE Trans. Mobile Comput.*, vol. 7, no. 8, pp. 915–930, Aug. 2008.
36. L. Le and E. Hossain, "A MAC protocol for opportunistic spectrum access in cognitive radio networks," in *Proc. IEEE Wireless Commun. Netw. Conf. (WCNC)*, Las Vegas, NV, USA, Mar.–Apr. 2008.
37. S. Huang, X. Liu, and Z. Ding, "Opportunistic spectrum access in cognitive radio networks," in *Proc. IEEE Intl. Conf. on Computer Commun. (INFOCOM)*, Phoenix, AZ, USA, Apr. 2008.
38. Y. Chen, Q. Zhao, and A. Swami, "Joint design and separation principle for opportunistic spectrum access in the presence of sensing errors," *IEEE Trans. Inf. Theory*, vol. 54, no. 5, pp. 2053–2071, May 2008.
39. Q. Zhao, B. Krishnamachari, and K. Liu, "On myopic sensing for multichannel opportunistic access: Structure, optimality, and performance," *IEEE Trans. Wireless Commun.*, vol. 7, no. 12, pp. 5431–5440, Dec. 2008.
40. L. Lai, H. E. Gamal, H. Jiang, and H. Poor, "Cognitive medium access: Exploration, exploitation and competition," *IEEE Trans. Mobile Comput.*, vol. 10, no. 2, pp. 239–253, Feb. 2011.
41. K. Liu, Q. Zhao, and B. Krishnamachari, "Dynamic multichannel access with imperfect channel state detection," *IEEE Trans. Signal Process.*, vol. 58, no. 5, pp. 2795–2808, May 2010.
42. K. W. Sung, S. L. Kim, and J. Zander, "Temporal spectrum sharing based on primary user activity prediction," *IEEE Trans. Wireless Commun.*, vol. 9, no. 12, pp. 3848–3855, Dec. 2010.
43. K. W. Choi and E. Hossain, "Opportunistic access to spectrum holes between packet bursts: A learning-based approach," *IEEE Trans. Wireless Commun.*, vol. 10, no. 8, pp. 2497–2509, Aug. 2011.
44. Q. Zhao, S. Geirhofer, L. Tong, and B. M. Sadler, "Opportunistic spectrum access via periodic channel sensing," *IEEE Trans. Signal Process.*, vol. 56, no. 2, pp. 785–796, Feb. 2008.
45. X. Liu, B. Krishnamachari, and H. Liu, "Channel selection in multi-channel opportunistic spectrum access networks with perfect sensing," in *Proc. IEEE Intl. Symp. New Frontiers in Dynamic Spectrum Access Networks (DySPAN)*, Singapore, Apr. 2010.
46. R. Smallwood and E. Sondik, "The optimal control of partially observable Markov processes over a finite horizon," *Operations Research*, vol. 21, no. 5, pp. 1071–1088, 1973.
47. E. Hossain, L. Le, N. Devroye, and M. Vu, "Cognitive radio: From theory to practical network engineering," in *New Directions in Wireless Communications Research*. Springer, 2009, pp. 251–289.
48. D. A. Berry and B. Fristedt, *Bandit problems: Sequential allocation of experiments*. London: Chapman and Hall, 1985.
49. M. Derakhshani and T. Le-Ngoc, "Learning-based opportunistic spectrum access with adaptive hopping transmission strategy,", *IEEE Trans. Wireless. Commun.*, vol. 11, no. 11, pp. 3957–3967, Nov. 2012.
50. M. Derakhshani and T. Le-Ngoc, "Distributed learning-based spectrum allocation with noisy observations in cognitive radio networks,", *IEEE Trans. Veh. Technol.*, vol. 63, no. 8, pp. 3715–3725, Oct. 2014.
51. M. Derakhshani and T. Le-Ngoc, "Intelligent CSMA-based opportunistic spectrum access: Competition and cooperation," in *Proc. IEEE Global Commun. Conf. (GLOBECOM)*, Anaheim, CA, USA, Dec. 2012.
52. M. Derakhshani and T. Le-Ngoc, "Adaptive access control of CSMA/CA in wireless LANs for throughput improvement," in *Proc. IEEE Global Commun. Conf. (GLOBECOM)*, Atlanta, GA, USA, Dec. 2013.

Chapter 3
Cognitive MAC Designs: Hopping Transmission Strategy

3.1 Introduction

As previously discussed in Chap. 1, the development and implementation of cognitive MAC protocols in OSA networks involve challenges that are not present in the existing radio systems. More specifically, in an OSA network, each SU needs to identify spectrum opportunities, coordinate their sharing with other competing SUs, and release them when they are acquired by PUs. Since PUs have higher priority to access the spectrum, two conflicting objectives arise in designing a cognitive MAC protocol for SUs. In particular, the opportunistic spectrum utilization of SUs needs to be maximized, while possible collisions between SUs to PUs must be kept limited [1,2]. Considering such conflicts between SUs and PUs, in this chapter, the aim is to develop an efficient MAC protocol for the SUs.

Recently, the cognitive MAC design in OSA networks has received considerable attention [3–8]. There are several MAC protocols presented in the literature that allow SUs to choose a frequency slot (or channel) to sense and, if available, access for an entire transmission duration. These strategies assume that both PUs and SUs have the same transmission time-slot structure [1, 9–12]. Such assumption of synchronous slotted transmission structure between PU and SU networks is not a sensible assumption since it needs good coordination in time between the PU and SU networks. Without assuming synchronization between PUs and SUs, PU activity in a channel with respect to SUs is dynamic and can be represented as a continuous-time ON/OFF random process (e.g., [13]). As a result, even if an idle channel is perfectly detected and used by a SU, a collision can still occur since the PU may return and reoccupy that idle channel at any time during the SU transmission.

To deal with the aforementioned issue, in this chapter, an adaptive hopping transmission strategy is proposed for SUs. In particular, this transmission strategy aims to reduce the effects of collision between PUs and SUs due to PU return in consideration of the PU dynamics assuming no time coordination between SUs and PUs. In the proposed scheme, instead of sensing and selecting one idle channel for the entire transmission as in previous studies, the SU accesses multiple idle channels,

© The Author(s) 2014   35
M. Derakhshani, T. Le-Ngoc, *Cognitive MAC Designs for OSA Networks*,
SpringerBriefs in Electrical and Computer Engineering, DOI 10.1007/978-3-319-12649-4_3

each with a different sojourn time (called activity factor). In this case, possible PU return in a channel may destroy only a small fraction of the SU transmission that can be recovered by erasure-correction coding to improve the SU transmission performance. Taking into account spectrum sharing among SUs, the dynamic PU activity, and also channel characteristics, the SU activity factor optimization problem is formulated to maximize the overall SU throughput. Subsequently, optimal MAC strategies are developed for SUs, based on the Lagrange dual decomposition method.

In addition, the proposed dual decomposition method—that provides a global solution to the optimization problem with affordable complexity—gives rise to the realization of distributed implementation. In other words, the optimal MAC strategy can be potentially performed in a distributed manner by each SU, provided that the knowledge of other SUs' activity factors are available. Such information can be obtained either with the aid of a central coordinator or by exchange of overhead information among SUs which may cause complexity and result in an un-scalable system. Consequently, we present a random-access-based MAC protocol in which each SU adjusts its activity factors independently by learning from the locally available information. In the proposed MAC protocol, each SU learns to respond optimally to its environment and adapts its activity factors to the optimal values over time. This fully distributed learning-based MAC protocol distinguishes this work in coordinating spectrum access among SUs from the previously proposed channel assignment schemes for SUs with support of a central controller (e.g., [14]) or exploiting a common control channel (e.g., [15]).

The implementation of the proposed learning-based MAC depends on the estimation of the sum of activity factors of all SUs in each channel. Since estimation with limited samples suffers from random errors, the proposed learning-based MAC protocol can be cast in the framework of stochastic gradient descent optimization. We analytically investigate its convergence and convergence rate to characterize its asymptotic behavior and efficiency. The main difficulty in analyzing its convergence properties lies in the fact that the estimation errors are biased.

3.2 System Model and Problem Formulation

Consider a frequency band licensed to PUs, which is divided into N_p non-overlapping frequency slots (or channels), each with bandwidth B^i. The frequency channels are chosen such that multipath fading can be considered constant (i.e., flat fading) over each channel (e.g., orthogonal frequency-division multiplexing (OFDM) narrowband subcarriers). Furthermore, an ad-hoc secondary network is considered with N_s SUs looking for temporal spectrum opportunities in these N_p channels. SUs are assumed to follow a slotted transmission scheme. Each time-slot of equal duration T consists of two periods: *sensing* of duration τ and *transmission* of duration $(T - \tau)$. In this work, we assume sufficiently accurate sensing with negligible PU miss-detection. Perfect sensing could be a sensible assumption in certain scenarios, e.g., applications in which SUs are located inside the service area of the PU transmitter (e.g., [13]).

Let $\mathcal{N}_a := \{1,\ldots,N_a\}$ denote the set of N_a channels that are detected idle at the beginning of each time-slot, and hence, can be utilized by N_s SUs. It is assumed that SUs are able to perform full spectrum sensing of N_p channels to find the idle channels in each time-slot. Furthermore, let C_k^i reflect channel quality (i.e., bits per second) for the SU k in the channel i and $g_{k,k}^i$ denote the power gain of the SU link k in the channel i. Consider a block flat fading situation in which $g_{k,k}^i$ remains unchanged during a given time-slot but independently varies from one time-slot to another. Thus, the transmission capacity of SU k in channel i is $B^i \log(1 + P_k^i g_{k,k}^i / n_k^i)$ where P_k^i and n_k^i represent the signal power and the noise power for the SU k in the channel i, respectively. In this work, without loss of generality, we consider the transmission capacity of SU k in channel i as a measure for C_k^i.

In general, the PU can use a time frame different from that of SUs. From the SU viewpoint, the PU activity in a given channel can be modeled as a two-state continuous-time random process with the ON (1) and OFF (0) states representing the *busy* and *idle* periods of the PU, respectively.

Given that channel i is detected idle in the sensing slot, the conditional probability that it becomes busy t seconds later is denoted by $\pi_{01}^i(t)$ which is an increasing function of time. Note that, by properly designing transmission slot duration (i.e., $T - \tau$), the increase of instantaneous PU return probability over a transmission slot can be kept negligible. Subsequently, the probability—that the PU reoccupies channel i in a transmission slot given that channel i is detected idle in the sensing slot at the beginning of that time-slot—is defined as the average of instantaneous probability of PU return over the whole transmission time, i.e., $\alpha_i = \frac{1}{T-\tau} \int_{t=0}^{T-\tau} \pi_{01}^i(t) dt$. As a result, α_i is a time-independent function, however, it is still a function of spectrum usage statistics of the PUs. On one hand, the PU return causes collision for the PU. In order to protect the PU transmission quality, α_i can be kept smaller than a required level by designing proper transmission duration (i.e., $T - \tau$). For instance, in [16, 17], it is demonstrated that by selecting a suitable spectrum sensing time and a transmission time, the SU can maximize its achievable throughput under the constraint that PUs are adequately protected.

On the other hand, the PU return may destroy the entire on-going SU transmission in the channel i. To avoid such a serious data loss due to the PU return, we propose an adaptive transmission strategy for SUs in which a SU dynamically hops over *multiple idle* channels, each with an unequal sojourn time (called activity factor) to be determined, so that possible PU return in a channel may destroy only a small fraction of the SU transmission that can be recovered by erasure-correction coding. Let $\beta_k^i (0 \leq \beta_k^i \leq 1)$ denote the activity factor of SU k in channel $i \in \mathcal{N}_a$ during a transmission slot. The activity-factor matrix of all β_k^i of all N_s SUs in N_a idle channels can be presented as $\boldsymbol{\beta} = \begin{bmatrix} \beta_1^1 & \cdots & \beta_1^{N_a} \\ \vdots & \ddots & \vdots \\ \beta_{N_s}^1 & \cdots & \beta_{N_s}^{N_a} \end{bmatrix}$.

Note that β_k^i is restricted to a binary value in previous studies [1, 9–12, 18, 19], such that $\beta_k^i = 0$ expresses that the SU k does not transmit in the channel i and $\beta_k^i = 1$ represents that the SU k transmits in the channel i for the entire transmission

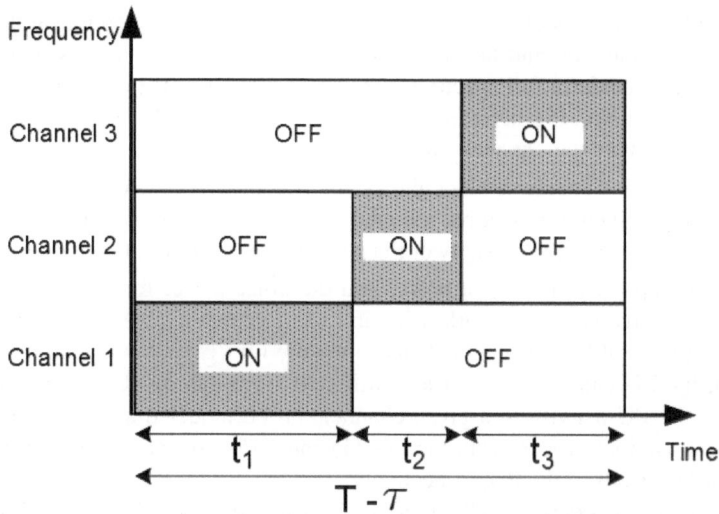

Fig. 3.1 An example of a SU activity in three idle channels with different activity factors during a transmission slot

slot. Therefore, the proposed transmission scheme with $0 \le \beta_k^i \le 1$ generalizes the existing schemes. Figure 3.1 illustrates an example of a SU activity according to the proposed hopping transmission strategy during a single transmission slot. In this example, assuming that the SU needs one channel for its transmission and $N_a = 3$, the activity factors of the SU in the different channels can be calculated as $\beta_k^i = \frac{t_i}{T-\tau}, i = 1,2,3$.

To coordinate spectrum access among SUs, it is assumed that SUs share an idle channel orthogonally in time domain, and hence, there is no mutual interference between SUs. Since one SU can exclusively use an idle channel at a certain point of time during each transmission slot, it is needed to assure that the total activity factors of different SUs in each idle channel remains smaller than 1. This constraint guarantees possible orthogonal sharing in time for each idle channel. Thus, $\sum_{k=1}^{N_s} \beta_k^i \le 1, i = 1,\ldots,N_a$.

Consider that the SU k needs a fixed number of channels, R_k, for its transmission during a single transmission slot. It follows that the sum of all activity factors of each SU k over all idle channels must equal its required number of channels, i.e., $\sum_{i=1}^{N_a} \beta_k^i = R_k$.

The normalized transmission rate of SU k in channel i is $\beta_k^i C_k^i$ since it transmits partially with activity factor β_k^i. Taking into account the possible loss in SU transmission due to PU return in each idle channel i with probability of α_i, the throughput of SU k in the idle channel i is defined as its *successful* transmission rate

$$f_k^i = \beta_k^i C_k^i \left(1 - \beta_k^i \alpha_i\right) \tag{3.1}$$

where $\beta_k^i \alpha_i$ expresses the probability that the SU k experiences transmission loss due to the PU return. In other words, the PU return probabilities, α_i, enable modeling the effect of dynamic PU activities on SU performance.

The goal of this work is to develop a resource allocation scheme that determines the optimal activity factors for N_s SUs in N_a idle channels in a single transmission slot to maximize the overall throughput of all SUs (i.e., $f = \sum_{k=1}^{N_s} \sum_{i=1}^{N_a} f_k^i$) under constraints of ensuring possible orthogonal time sharing in each idle channel (i.e., $\sum_{k=1}^{N_s} \beta_k^i \le 1$) and fixed channel requirements for SUs (i.e., $\sum_{i=1}^{N_a} \beta_k^i = R_k$).

3.3 Optimal Hopping-Based MAC Design

In this section, we study optimization problems which offer optimum activity factors considering the adaptive hopping-based transmission strategy for SUs. First, the activity factor optimization problem is developed for a single SU to find the best solution for each SU without consideration of the spectrum sharing with the other SUs. Then, the activity factor optimization problem formulation and algorithm development are discussed for multiple SUs.

3.3.1 Single-User

First, the activity factor optimization problem is considered for a single SU. The index of SU k is set to 1 without loss of generality in this section. In this case, the optimization problem can be formulated as

$$\max_{\boldsymbol{\beta}_1} \sum_{i=1}^{N_a} \beta_1^i C_1^i \left(1 - \beta_1^i \alpha_i\right) \tag{3.2a}$$

$$\text{subject to} \sum_{i=1}^{N_a} \beta_1^i = R_1 \tag{3.2b}$$

$$0 \le \beta_1^i \le 1, \ i = 1, \dots, N_a. \tag{3.2c}$$

Note that the objective function in (3.2a) expresses the overall SU throughput over all idle channels and R_1 in (3.2b) represents the channel requirement of the SU. Based on (3.1), the overall throughput of the SU k over all idle channels can be simply presented by $\sum_{i=1}^{N_a} f_k^i$ due to independent activity of the SU k over different idle channels. Note that there are no specific assumptions on the joint distribution of the PU activities to formulate the overall throughput in (3.2a).

Since the objective function in (3.2a) is concave (i.e., the negative of objective function is convex) and the equality constraint in (3.2b) is a linear function of $\boldsymbol{\beta}_1 = \left[\beta_1^1, \dots, \beta_1^{N_a}\right]$, this problem is a convex optimization with a coupling constraint in (3.2b). A dual decomposition is an appropriate approach to solve the convex problem with zero duality gap when the problem has coupling constraints. By relaxing the coupling constraints, the optimization problem decouples into several sub-problems [20]. To relax the coupling constraint in (3.2b), it makes sense to write the Lagrangian function of (3.2) as

$$\mathscr{L}(\lambda, \boldsymbol{\beta}_1) = \sum_{i=1}^{N_a} \beta_1^i C_1^i (1 - \beta_1^i \alpha_i) - \lambda \left(\sum_{i=1}^{N_a} \beta_1^i - R_1 \right) \tag{3.3}$$

where λ denotes the Lagrange multiplier corresponding to (3.2b). The optimization problem can be separated into two levels of optimization. At the lower level, there are N_a sub-problems for each idle channel with Lagrangian $\mathscr{L}(\lambda, \beta_1^i) = \beta_1^i C_1^i (1 - \beta_1^i \alpha_i) - \lambda \beta_1^i$ assuming λ is fixed. Subsequently, it can be solved by writing the Karush-Kuhn-Tucker (KKT) conditions

$$\frac{\partial \mathscr{L}(\lambda, \beta_1^i)}{\partial \beta_1^i} = C_1^i - 2C_1^i \beta_1^i \alpha_i - \lambda = 0. \tag{3.4}$$

Then, considering $0 \leq \beta_1^i \leq 1$, we obtain $\beta_1^i(\lambda) = \left[\frac{C_1^i - \lambda}{2C_1^i \alpha_i} \right]_0^1$ where $[x]_a^b = \min(b, \max(a, x))$. At the higher level, there is the master dual problem responsible for updating the dual variable (i.e., λ) by solving the dual problem. Then, λ can be found iteratively with the help of the following gradient method

$$\lambda^{n+1} = \lambda^n + \varepsilon \left(\sum_{i=1}^{N_a} \beta_1^i(\lambda^n) - R_1 \right) \tag{3.5}$$

where n is the iteration index, ε is a sufficiently small positive step-size. Since the proposed transmission strategy generalizes the existing approaches, the non-binary optimal activity factors prove the benefit of the proposed approach over the existing schemes in offering throughput improvement. This can be explained by the fact that using more idle channels, each of them for a fraction of transmission slot, helps the SU to decrease transmission loss due to the collision caused by the PU return.

Assuming that there is a λ such that $0 \leq \frac{C_1^i - \lambda}{2C_1^i \alpha_i} \leq 1$, $\forall i \in \mathcal{N}_a$, then, $\beta_1^i = \frac{C_1^i - \lambda}{2C_1^i \alpha_i}$, and, according to the equality constraint (3.2b),

$$\sum_{i=1}^{N_a} \frac{C_1^i - \lambda}{2C_1^i \alpha_i} = R_1 \Rightarrow \lambda = \frac{\sum_{i=1}^{N_a} \frac{1}{2\alpha_i} - R_1}{\sum_{i=1}^{N_a} \frac{1}{2C_1^i \alpha_i}}. \tag{3.6}$$

Therefore, the optimum activity factor in the channel i can be represented as

$$\beta_1^i = \frac{R_1 + C_1^i \sum_{i=1}^{N_a} \frac{1}{2C_1^i \alpha_i} - \sum_{i=1}^{N_a} \frac{1}{2\alpha_i}}{2C_1^i \alpha_i \sum_{i=1}^{N_a} \frac{1}{2C_1^i \alpha_i}}. \tag{3.7}$$

3.3.2 Multi-User

Considering multiple SUs in a secondary network, the problem of spectrum sharing affects the MAC design. To coordinate the spectrum access among SUs, it is assumed that SUs share an idle channel orthogonally in time domain, and hence, there is no mutual interference between SUs. The goal is to maximize the overall throughput of SUs (i.e., $\sum_{k=1}^{N_s} \sum_{i=1}^{N_a} f_k^i$) under constraints of ensuring possible orthogonal time sharing in each idle channel and fixed channel requirements for SUs. More specifically, the optimization problem can be formulated as

$$\max_{\beta} \sum_{k=1}^{N_s} \sum_{i=1}^{N_a} \beta_k^i C_k^i \left(1 - \beta_k^i \alpha_i\right) \tag{3.8a}$$

$$\text{subject to } \sum_{k=1}^{N_s} \beta_k^i \le 1, \ i = 1, \ldots, N_a \tag{3.8b}$$

$$\sum_{i=1}^{N_a} \beta_k^i = R_k, \ k = 1, \ldots, N_s \tag{3.8c}$$

$$0 \le \beta_k^i \le 1, \ i = 1, \ldots, N_a, \ k = 1, \ldots, N_s. \tag{3.8d}$$

It is worth mentioning that the feasibility criterion for the above optimization problem is $\sum_{k=1}^{N_s} R_k \le N_a$. In other words, when the demand of all SUs combined (i.e., $\sum_{k=1}^{N_s} R_k$) is larger than the number of accessible idle channels (i.e., N_a), it is not possible to select a matrix of activity factors (i.e., β) which satisfies all constraints.

This problem is a convex optimization with coupling constraints in (3.8b) and (3.8c). In order to decouple this problem, first, a dual decomposition with respect to (3.8b) and then, for the sub-problem, another dual decomposition with respect to (3.8c) are applied. This offers a two-level optimization decomposition including a master dual problem, a secondary master dual problem, and the sub-problems [20]. By taking a relaxation of the coupling constraints in (3.8b), the Lagrangian function of the optimization problem in (3.8) becomes

$$\max_{\beta} \sum_{k=1}^{N_s} \sum_{i=1}^{N_a} \beta_k^i C_k^i \left(1 - \beta_k^i \alpha_i\right) - \sum_{i=1}^{N_a} \mu_i \left(\sum_{k=1}^{N_s} \beta_k^i - 1\right) \tag{3.9}$$

where $\mu_i \ge 0$ is the Lagrange multiplier associated with (3.8b). Therefore, at the lower level, for fixed μ_i, there are N_s sub-problems with the following objective function for each SU

$$\max_{\boldsymbol{\beta}_k} \sum_{i=1}^{N_a} \beta_k^i C_k^i \left(1 - \beta_k^i \alpha_i\right) - \sum_{i=1}^{N_a} \mu_i \beta_k^i \qquad (3.10a)$$

$$\text{subject to } \sum_{i=1}^{N_a} \beta_k^i = R_k \qquad (3.10b)$$

$$0 \le \beta_k^i \le 1, \; i = 1, \ldots, N_a \qquad (3.10c)$$

where $\boldsymbol{\beta}_k = \left[\beta_k^1, \ldots, \beta_k^{N_a}\right]$. The Lagrangian function of (3.10) for fixed $\boldsymbol{\mu} = [\mu_1, \ldots, \mu_{N_a}]$ is

$$\mathcal{L}\left(\lambda_k, \boldsymbol{\mu}, \boldsymbol{\beta}_k\right) = \sum_{i=1}^{N_a} \beta_k^i C_k^i \left(1 - \beta_k^i \alpha_i\right)$$
$$- \sum_{i=1}^{N_a} \mu_i \beta_k^i - \lambda_k \left(\sum_{i=1}^{N_a} \beta_k^i - R_k\right) \qquad (3.11)$$

where λ_k is the Lagrange multiplier associated with (3.10b). This optimization problem can also be separated into two levels of optimization. At the lower level, there are N_a sub-problems for each idle channel with Lagrangian $\mathcal{L}\left(\lambda_k, \mu_i, \beta_k\right) = \beta_k^i C_k^i \left(1 - \beta_k^i \alpha_i\right) - \mu_i \beta_k^i - \lambda_k \beta_k^i$ assuming λ_k is fixed. Subsequently, it can be solved by writing the KKT conditions

$$\frac{\partial \mathcal{L}\left(\lambda_k, \mu_i, \beta_k\right)}{\partial \beta_k^i} = C_k^i - 2C_k^i \beta_k^i \alpha_i - \mu_i - \lambda_k = 0. \qquad (3.12)$$

Then, considering $0 \le \beta_k^i \le 1$,

$$\beta_k^i (\lambda_k, \mu_i) = \left[\frac{C_k^i - \mu_i - \lambda_k}{2C_k^i \alpha_i}\right]_0^1 \qquad (3.13)$$

where $[x]_a^b = \min\left(b, \max\left(a, x\right)\right)$. At the higher level, there exists the secondary master dual problem responsible for updating the dual variable λ_k by solving the secondary dual problem. Then, λ_k can be found iteratively with the help of the following gradient method

$$\lambda_k^{n+1} = \lambda_k^n + \nu_n \left(\sum_{i=1}^{N_a} \beta_k^i (\lambda_k^n, \mu_i^n) - R_k\right) \qquad (3.14)$$

where n is the iteration index and $\nu_n > 0$ is a sequence of scalar step-sizes. λ_k (called SU k price) is updated until satisfying the channel requirement for the SU k. Then, at the highest level, μ_i (called channel i price) is updated with a slower rate until satisfying $\sum_{k=1}^{N_s} \beta_k^i \le 1$. The μ_i can be found iteratively with the help of gradient approach. The following gradient method can be used

---

Algorithm 3.1: Optimal MAC

---

1 Set $n = 0$, μ_i^0 equal to some non-negative value for all i and λ_k^0 equal to some value for all k.
2 β_k^i is computed for all k and i according to (3.13).
3 Fast price updating: each SU price λ_k is updated with the gradient iteration (3.14).
4 Set $n \leftarrow n+1$ and go to step 2 (until satisfying termination criterion).
5 Slow price updating: each channel price μ_i is updated according to the gradient iteration (3.15).
6 Go to step 2 (until satisfying termination criterion).

$$\mu_i^{n+1} = \left[\mu_i^n + \gamma_n \left(\sum_{k=1}^{N_s} \beta_k^i(\lambda_k^n, \mu_i^n) - 1\right)\right]^+ \tag{3.15}$$

where $\gamma_n > 0$ is a sequence of scalar step-sizes. Also, $[.]^+$ denotes the projection onto the non-negative space. The dual variables will converge to the dual optimums as long as v_n and γ_n are chosen to be sufficiently small. Some popular choices for v_n and γ_n include constant values such as $\delta > 0$ or diminishing functions of time such as an^{-c}, where $a > 0$ and $0 < c \leq 1$ are constant scalars. Since the duality gap is zero due to the convex optimization, the primal variable β_k^i will also converge to the optimal value as well.

Based on the above derivations, the optimal MAC protocol is presented in Algorithm 3.1, with an inner loop to update the SU prices (i.e., λ_k) and an outer loop to update the channel prices (i.e., μ_i). Algorithm 3.1 needs to be performed at the beginning of every time-slot of duration T to find the optimal activity factors of SUs. However, considering slowly varying statistics of PU activity and slow fading, the frequency that optimization problem needs to be solved will decrease. In Algorithm 3.1, it is noted that a central coordinator is responsible to manage the sequence of access of different SUs in different channels based on their optimal activity factors.

Apparently, Algorithm 3.1 (i.e., the optimal MAC) can be potentially performed in a distributed manner by each SU. However, from (3.15), it is obvious that each SU needs to know the sum of activity factors of all N_s SUs in the idle channel i, $\beta^i = \sum_{k=1}^{N_s} \beta_k^i(t)$, to update the channel price μ_i. Such information can be obtained with the aid of a central coordinator or heavy exchange of overhead information which causes complexity and results in an un-scalable system. It is thus crucial that SUs learn this information to update channel prices independently which will be used to adjust their activity factors. It is noted that perfect knowledge of channel gains are required to update β_k^i based on (3.13). However, in the potential distributed implementation, each SU needs only the channel state information of its own link (i.e., the channel state information of other SU links are not required).

3.4 Learning-Based Distributed MAC Design via Adaptive Carrier Sensing

In this section, to enable distributed implementation of the optimal MAC protocol, an adaptive carrier sensing multiple access (CSMA) scheme is devised as a decentralized mechanism to access an idle channel for SUs based on their activity factors. Then, we discuss how to use the capturing status feedbacks of the proposed adaptive CSMA scheme to estimate the sum of activity factors of all SUs in each idle channel, which will be employed to update the channel prices (i.e., μ_i). Subsequently, the channel price updating is modeled as a stochastic gradient descent method considering erroneous estimations of β^i.

It is proposed that SUs share idle channels by means of an adaptive CSMA based on their activity factors without a central coordinator. In the adaptive CSMA scheme, each transmission slot is divided into S equal sub-slots with length $\frac{T-\tau}{S}$, labeled $t_1, \ldots, t_S$. In each sub-slot t_j in channel i, the SU k performs the following steps:

1- Generate a Bernoulli random variable $x_k^i(t_j)$ with the success probability β_k^i to be determined. If $x_k^i(t_j) = 0$, the SU k will not transmit in the subslot t_j. If $x_k^i(t_j) = 1$, the SU k will proceed to the next step.
2- Generate a backoff time $W_k^i(t_j)$ according to a uniform distribution in the interval $(0, W)$.
3- After expiry of the backoff time, sense the channel i, if it is idle, transmit.

In this proposed adaptive CSMA scheme, one SU with the smallest backoff time among the SUs who compete for the same sub-slot (i.e., $x_k^i(t_j) = 1$) will succeed and transmit in this sub-slot. Let $y_k^i(t_j)$ be a binary random variable representing the capturing status: $y_k^i(t_j) = 1$ if the SU k captures the channel i in sub-slot t_j; otherwise, $y_k^i(t_j) = 0$.

Assuming that each SU k keeps track of its capturing status feedbacks, $y_k^i(t_j)$, the *achieved* activity factor of SU k in the channel i (i.e., the average time proportion of a transmission slot that an SU *successfully* occupies the channel i, given the competition among SUs) can be obtained as

$$\bar{\beta}_k^i = \frac{\mathrm{E}\left[\sum_{j=1}^{S} x_k^i(t_j) y_k^i(t_j)\right]}{S}. \tag{3.16}$$

Then,

$$\begin{aligned}
\bar{\beta}_k^i &= \mathrm{Prob}\left[y_k^i(t_j) = 1, x_k^i(t_j) = 1\right] \\
&= \beta_k^i \mathrm{Prob}\left[y_k^i(t_j) = 1 | x_k^i(t_j) = 1\right]
\end{aligned} \tag{3.17}$$

where $\beta_k^i = \mathrm{Prob}[x_k^i(t_j) = 1]$ represents the *intended* activity factor of SU k in channel i. The probability of getting the smallest backoff time to capture the channel is inversely proportional to the number of SUs actively competing to capture the same channel, i.e.,

$$\text{Prob}\left[y_k^i(t_j) = 1 | x_k^i(t_j) = 1\right] = \text{E}\left[\frac{1}{1 + \sum_{k=1, \bar{k} \neq k}^{N_s} x_k^i(t_j)}\right]. \qquad (3.18)$$

Applying the Jensen's inequality, $\text{Prob}\left[y_k^i(t_j) = 1 | x_k^i(t_j) = 1\right] \geq \frac{1}{1 + \sum_{k=1, \bar{k} \neq k}^{N_s} \beta_k^i}$. In order to find a closed-form expression, $\text{Prob}\left[y_k^i(t_j) = 1 | x_k^i(t_j) = 1\right]$ is approximated with the lower-bound, i.e.,

$$\bar{\beta}_k^i \simeq \frac{\beta_k^i}{1 + \sum_{\bar{k}=1, \bar{k} \neq k}^{N_s} \beta_{\bar{k}}^i}. \qquad (3.19)$$

Note that (3.19) shows that the *achieved* activity factor of each SU using the proposed CSMA scheme depends on the *intended* activity factor of other SUs. However, by keeping $\sum_{k=1}^{N_s} \beta_k^i \leq 1$, it can be guaranteed that the *achieved* activity factor is always larger than the half of *intended* activity factor (i.e., $\bar{\beta}_k^i \geq \frac{\beta_k^i}{2}$). Thus, to reduce computational complexity, here, the similar utility function is considered as in (3.8), along with $\sum_{k=1}^{N_s} \beta_k^i \leq 1$ to manage competition among SUs. Note that this helps to alleviate *achieved* activity factor (i.e., $\bar{\beta}_k^i$) degradation due to the crowding effects.

Apparently, in Algorithm 3.1, each SU k needs the sum of the activity factors of all N_s SUs in channel i, $\sum_{k=1}^{N_s} \beta_k^i$, to update the channel price μ_i in (3.15). From (3.19), the SU k can estimate $\sum_{\bar{k}=1, \bar{k} \neq k}^{N_s} \beta_{\bar{k}}^i \simeq \left(\frac{\beta_k^i}{\bar{\beta}_k^i} - 1\right)$ where its *achieved* activity factor $\bar{\beta}_k^i$ can be updated after a window of S' sub-slots, based on the available observations of $y_k^i(t_l)$, $l = f, \dots, f + S'$ where $0 \leq f \leq S - S'$, as $\bar{\beta}_k^i = \sum_{l=f}^{f+S'} y_k^i(t_l)/S'$. Hence, $\beta^i = \sum_{k=1}^{N_s} \beta_k^i$ can be updated after each S' sub-slots as

$$\hat{\beta}^i \simeq \beta_k^i + \frac{S' \cdot \beta_k^i}{\sum_{l=f}^{f+S'} y_k^i(t_l)} - 1. \qquad (3.20)$$

Note that if $\beta_k^i = 0$, the SU k can estimate $\sum_{\bar{k}=1, \bar{k} \neq k}^{N_s} \beta_{\bar{k}}^i$ by keeping track of its capturing status feedbacks, $y_k^i(t_j)$, while assuming a virtual activity factor $\tilde{\beta}_k^i = 1$, i.e., $\sum_{\bar{k}=1, \bar{k} \neq k}^{N_s} \beta_{\bar{k}}^i \simeq \left(\frac{\tilde{\beta}_k^i}{\bar{\beta}_k^i} - 1\right)$. In other words, the SU k performs the proposed CSMA scheme with $\tilde{\beta}_k^i = 1$ and sets $y_k^i(t_j) = 1$ if it achieves the smallest back-off time. However, since the actual activity factor is zero, the SU k does not transmit in order to avoid affecting other SUs' estimation process.

Since the estimator $\hat{\beta}^i$ is a non-linear function of $\bar{\beta}_k^i = \frac{\sum_{l=f}^{f+S'} y_k^i(t_l)}{S'}$, it can be proved that the bias and the variance of $\hat{\beta}^i$ are of $\mathcal{O}((S')^{-1})$ [21]. In other words, considering $\hat{\beta}^i = \beta^i + w = \sum_{k=1}^{N_s} \beta_k^i + w$ where w denotes the random error, we have

---
Algorithm 3.2: Learning-based MAC
---
1 Set $n = 0$, μ_i^0 equal to some non-negative value for all i and λ_k^0 equal to some value.
2 Each SU k computes β_k^i for all i according to (3.13).
3 Fast price updating: each SU k updates λ_k price with the gradient iteration (3.14).
4 Set $n \leftarrow n+1$ and go to step 2 (until satisfying termination criterion).
5 According to the selected β_k^i in different channels, each SU k starts transmission for S' sub-slots based on the proposed CSMA procedure.
6 Slow price updating after each S' sub-slots: Each SU k updates $\sum_{k=1}^{N_s} \beta_k^i$ according to (3.20) and μ_i according to the gradient iteration (3.22).
7 Go to step 2 (until satisfying the termination criterion).

$$E[w] \le K_e \left(S'\right)^{-1} \text{ and } \mathrm{var}[w] \le K_v \left(S'\right)^{-1}. \tag{3.21}$$

From (3.15), in the optimal MAC protocol, the channel price μ_i is updated using a gradient method with $\nabla g(\mu_i) = 1 - \sum_{k=1}^{N_s} \beta_k^i = 1 - \beta^i$. However, in the learning-based MAC, only noisy measurements of $\nabla g(\mu_i)$ are available for SUs. Based on (3.15), using the estimation of $\beta^i = \sum_{k=1}^{N_s} \beta_k^i$, channel price μ_i can be updated independently by each SU as

$$\mu_i^{n+1} = \left[\mu_i^n + \gamma_n(\hat{\beta}^i - 1)\right]^+$$
$$= [\mu_i^n + \gamma_n(-\nabla g(\mu_i^n) + w_n)]^+. \tag{3.22}$$

Comparing to (3.15), the channel price updating in (3.22) involves stochastic errors (i.e., w_n), and hence, the popular stochastic gradient descent method [22, 23] is exploited to study the convergence of the proposed learning-based MAC protocol.

Based on the above derivations, the fully-distributed MAC protocol is presented in Algorithm 3.2, which can be separately performed by each SU to determine its optimal activity factors. Each SU needs to perform Algorithm 3.2 at the beginning of every time-slot of duration T to update its own activity factors iteratively until convergence to the optimal values. It is noted that the average time scale of SU activity (i.e., being on and off) is a sub-slot in which a successful transmission could happen. Thus, on one hand, a sub-slot has to be large enough to accommodate a transmission with all overheads and collision avoidance signaling. On the other hand, it should be short enough that S sub-slots (i.e., the whole transmission duration) can be smaller than the coherence time since a block flat fading situation is considered.

3.5 Convergence Analysis

In this section, our goal is to prove that if each SU autonomously deploys (3.22) to update the channel prices, then the network performance converges to the optimal value. Thus, we investigate the convergence and convergence rate, which characterize the asymptotic behavior and efficiency of the proposed MAC strategy while it is formulated as a stochastic gradient descent method. To study the convergence, the following assumptions are adopted on the step size γ_n and learning window size S'.

Assumption 3.1 γ_n *is assumed as a deterministic positive step size satisfying* $\gamma_n > 0, \gamma_n \to 0, \sum_{n=0}^{\infty} \gamma_n \to \infty, \sum_{n=0}^{\infty} \gamma_n^2 < \infty.$

Note that Assumption 3.1 is widely used in the stochastic gradient search literature [23–25]. This is because, these conditions ensure a balance to have the step size decay neither too slow nor too fast. In particular, the step size should approach zero sufficiently slow $(\gamma_n \to 0, \sum_{n=0}^{\infty} \gamma_n \to \infty)$ to avoid false convergence while approaching zero at a sufficiently fast rate $(\gamma_n \to 0, \sum_{n=0}^{\infty} \gamma_n^2 < \infty)$ to diminish the noise effects as the iteration gets close to the optimal solution. A common generalization of step size sequence is $\gamma_n = an^{-c}$ for $a > 0$ and $0.5 < c \leq 1$ [23].

Assumption 3.2 *The size of learning window (i.e., S') is assumed as an increasing function of time $S' = K_b n^b$, $K_b \geq 1$, $b > 0$ such that $b + c > 1$. The condition $b + c > 1$ forces sufficiently fast decay of estimation bias.*

Remark 3.1 *Based on (3.21), Assumption 3.2 guarantees that the bias and variance of gradient estimator are diminishing functions of time. Specifically, $\mathrm{E}[w_n] \leq K_e n^{-b}$ and $\mathrm{var}[w_n] \leq K_v n^{-b}$.*

Remark 3.2 $\nabla g(\mu_i)$ *has Lipschitz continuity, i.e., there exists a positive real constant D_0 such that $|\nabla g(\mu_i) - \nabla g(\mu_i')| \leq D_0 |\mu_i - \mu_i'|$ for all positive μ_i and μ_i'. This can be explained by the fact that $\nabla g(\mu_i)$ is a continuous and linear function of μ_i based on (3.13), and also bounded (i.e., $1 - N_s \leq \nabla g(\mu_i) = 1 - \sum_{k=1}^{N_s} \beta_k^i \leq 1$).*

In the following proposition, we establish the convergence of $\mu_i^{n+1} = \mu_i^n + \gamma_n(-\nabla g(\mu_i^n) + w_n)$ which guarantees the convergence of (3.22), and hence, the activity factors of all SUs. This result is an extension of Theorem 1 in [26, p. 51] in consideration of biased errors.

Proposition 3.1 *(Convergence with probability of 1) Under Assumption 3.1 and Assumption 3.2, the sequence $\{\mu_i^n\}$ converges to the optimal value with probability of 1.*

Proof: From (3.22),

$$\left|\mu_i^{n+1} - \mu_i^*\right|^2 = |\mu_i^n - \mu_i^* + \gamma_n(-\nabla g(\mu_i^n) + w_n)|^2$$
$$= |\mu_i^n - \mu_i^*|^2 + 2\gamma_n(-\nabla g(\mu_i^n) + w_n)(\mu_i^n - \mu_i^*) + \gamma_n^2 |-\nabla g(\mu_i^n) + w_n|^2. \quad (3.23)$$

Since the dual objective function $g(\mu_i^n)$ is convex, $\nabla g(\mu_i^n)(\mu_i^n - \mu_i^*) \geq g(\mu_i^n) - g(\mu_i^*)$. Then,

$$\left|\mu_i^{n+1} - \mu_i^*\right|^2 \leq |\mu_i^n - \mu_i^*|^2 - 2\gamma_n(g(\mu_i^n) - g(\mu_i^*)) + 2\gamma_n w_n(\mu_i^n - \mu_i^*)$$
$$+ \gamma_n^2 |-\nabla g(\mu_i^n) + w_n|^2. \quad (3.24)$$

By adding and subtracting $\nabla g(\mu_i^*)$ in the last term of (3.24), since $|r+s|^2 \leq 2|r|^2 + 2|s|^2$ for any $r, s \in \mathbb{R}$,

$$\left|\mu_i^{n+1} - \mu_i^*\right|^2 \leq \left|\mu_i^n - \mu_i^*\right|^2 - 2\gamma_n \left(g\left(\mu_i^n\right) - g\left(\mu_i^*\right)\right) + 2\gamma_n w_n \left(\mu_i^n - \mu_i^*\right)$$
$$+ 2\gamma_n^2 \left|\nabla g\left(\mu_i^n\right) - \nabla g\left(\mu_i^*\right)\right|^2 + 2\gamma_n^2 \left|\nabla g\left(\mu_i^*\right) - w_n\right|^2. \qquad (3.25)$$

According to Remark 3.2, due to Lipschitz continuity of ∇g, we have

$$\left|\mu_i^{n+1} - \mu_i^*\right|^2 \leq (1 + 2D_0^2 \gamma_n^2) \left|\mu_i^n - \mu_i^*\right|^2 - 2\gamma_n \left(g\left(\mu_i^n\right) - g\left(\mu_i^*\right)\right) + 2\gamma_n w_n \left(\mu_i^n - \mu_i^*\right)$$
$$+ 2\gamma_n^2 \left|\nabla g\left(\mu_i^*\right) - w_n\right|^2. \qquad (3.26)$$

By taking the conditional expectation given $\mathscr{F}_n = \{\mu_i^0, \ldots, \mu_i^n\}$, we obtain

$$\mathrm{E}\left[\left|\mu_i^{n+1} - \mu_i^*\right|^2 \big| \mathscr{F}_n\right] \leq (1 + 2D_0^2 \gamma_n^2) \left|\mu_i^n - \mu_i^*\right|^2$$
$$- 2\gamma_n \left[g\left(\mu_i^n\right) - g\left(\mu_i^*\right) - \mathrm{E}[w_n]\left(\mu_i^n - \mu_i^*\right)\right] + 4\gamma_n^2 \left[\left|\nabla g\left(\mu_i^*\right)\right|^2 + \mathrm{E}[|w_n|^2]\right]. \qquad (3.27)$$

Based on Remark 3.1 in Chap. 3 and the fact that $\nabla g(\mu_i)$ is bounded, it is clear that $[|\nabla g(\mu_i^*)|^2 + \mathrm{E}[|w_n|^2]]$ is bounded, and hence, $\sum_{n=0}^{\infty} \gamma_n^2 \left[|\nabla g(\mu_i^*)|^2 + \mathrm{E}[|w_n|^2]\right] < \infty$ considering Assumption 3.1. Using Robbins-Siegmund Lemma in [26, p. 50], since (3.27) holds for all n, $\sum_{n=0}^{\infty} \gamma_n^2 < \infty$ and $\sum_{n=0}^{\infty} \gamma_n^2 \left[|\nabla g(\mu_i^*)|^2 + \mathrm{E}[|w_n|^2]\right] < \infty$, it can be concluded that the sequence $\{|\mu_i^n - \mu_i^*|^2\}$ is convergent and $\sum_{n=0}^{\infty} \gamma_n [g(\mu_i^n) - g(\mu_i^*) - \mathrm{E}[w_n](\mu_i^n - \mu_i^*)] < \infty$ with probability of 1. Since $\sum_{n=0}^{\infty} \gamma_n = \infty$ and $\lim_{n \to \infty} \mathrm{E}[w_n] = 0$ according to Remark 3.1, the later relation implies $\lim_{n \to \infty} g(\mu_i^n) = g(\mu_i^*)$ (i.e., $\lim_{n \to \infty} \nabla g(\mu_i^n) = 0$) with probability of 1. Therefore, $\{\mu_i^n\}$ converges to the μ_i^* with probability of 1. ∎

Remark 3.3 *If S' is fixed, $\lim_{n \to \infty} \left[(g(\mu_i^n) - g(\mu_i^*)) - \mathrm{E}[w_n](\mu_i^n - \mu_i^*)\right] = 0$. It implies that $\frac{g(\mu_i^n) - g(\mu_i^*)}{\mu_i^n - \mu_i^*} \to \mathrm{E}[w_n]$. Since $\{|\mu_i^n - \mu_i^*|^2\}$ is convergent, if S' is chosen sufficiently large (i.e., sufficiently small $\mathrm{E}[w_n]$), $g(\mu_i^n) - g(\mu_i^*)$ becomes sufficiently small. This explains that iterations (i.e., μ_i^n) converge in a neighborhood of the optimal value as moving in the direction of the optimal point.*

Remark 3.4 *For some positive real constants D_1 and D_2, and for all μ_i^n,*

$$D_1 \left|\mu_i^n - \mu_i^*\right| \leq \left|\nabla g(\mu_i^n) - \nabla g(\mu_i^*)\right| \leq D_2 \left|\mu_i^n - \mu_i^*\right|. \qquad (3.28)$$

This is because $\nabla g(\mu_i^n) = 1 - \sum_{k=1}^{N_s} \beta_k^i$ is a continuous and linear function of μ_i^n based on (3.13), and also bounded.

Next, our second main result is presented on the convergence rate of the learning-based algorithm. It reveals that the convergence rate is attached by how rapidly the learning window size increases and how fast the step size diminishes. The proof of the following proposition is built on the result for the convergence rate considering unbiased random errors in [27].

Proposition 3.2 *(Convergence rate) Let Assumption 3.1 and Assumption 3.2 hold. Then, for $0.5 < c < 1$, there exists an N_0 such that for all $n > N_0$, the following is true:*

$$E\left[|\mu_i^{n+1} - \mu_i^*|^2\right] \leq \frac{aD_3K_e^2q^{-1}}{1 - D_5q^{-1}}n^{-2b} + \frac{D_4q^{-1}}{1 - D_5q^{-1}}n^{-b-c}$$
$$+ \mathcal{O}\left(n^{-2b-1} + n^{-b-c-1}\right) \tag{3.29}$$

which implies $E\left[|\mu_i^{n+1} - \mu_i^*|^2\right] = \mathcal{O}\left(n^{-\min(2b,b+c)}\right)$. *And for $c = 1$,*

$$E\left[|\mu_i^{n+1} - \mu_i^*|^2\right] \leq D_8n^{-D_1a} + D_6n^{-2b} + D_7n^{-b-1} \tag{3.30}$$

which implies $E\left[|\mu_i^{n+1} - \mu_i^*|^2\right] = \mathcal{O}\left(n^{-\min(D_1a,2b,b+1)}\right)$. *Note that D_3, D_4, D_5, D_6, D_7, D_8 and q are positive real constants.*

Proof: By definition,

$$\mu_i^{n+1} - \mu_i^* = [\mu_i^n - \mu_i^* - \gamma_n \nabla g(\mu_i^n)] + \gamma_n w_n. \tag{3.31}$$

Squaring both sides of (3.31), taking the expected value and using $\gamma_n = an^{-c}$, we have

$$E\left[|\mu_i^{n+1} - \mu_i^*|^2\right] = E\left[|\mu_i^n - \mu_i^* - an^{-c}\nabla g(\mu_i^n)|^2\right] + a^2n^{-2c}E[|w_n|^2]$$
$$+ 2an^{-c}E\left[w_n\left(\mu_i^n - \mu_i^* - an^{-c}\nabla g(\mu_i^n)\right)\right]. \tag{3.32}$$

Using the inequality $xy \leq \frac{1}{2}(x^2 + y^2)$, the third term of (3.32) becomes

$$E\left[w_n(\mu_i^n - \mu_i^* - an^{-c}\nabla g(\mu_i^n))\right] = E[w_n]E[\mu_i^n - \mu_i^* - an^{-c}\nabla g(\mu_i^n)]$$
$$\leq \frac{1}{2}\left(E^2[w_n] + E^2[\mu_i^n - \mu_i^* - an^{-c}\nabla g(\mu_i^n)]\right)$$
$$\leq \frac{1}{2}\left(D_3K_e^2n^{-2b} + \frac{1}{D_3}E\left[|\mu_i^n - \mu_i^* - an^{-c}\nabla g(\mu_i^n)|^2\right]\right) \tag{3.33}$$

where D_3 is a positive real constant. According to Remark 3.1,

$$E[|w_n|^2] = \text{var}[w_n] + E^2[w_n] \leq K_v n^{-b} + K_e^2 n^{-2b}. \tag{3.34}$$

From (3.33) and (3.34), (3.32) can be written as

$$E\left[|\mu_i^{n+1} - \mu_i^*|^2\right] \leq (1 + \frac{a}{D_3}n^{-c})E\left[|\mu_i^n - \mu_i^* - an^{-c}\nabla g(\mu_i^n)|^2\right] + a^2K_v n^{-2c-b}$$
$$+ a^2K_e^2 n^{-2c-2b} + aD_3K_e^2 n^{-c-2b}. \tag{3.35}$$

Considering Remark 3.2,

$$\mathrm{E}\left[|\mu_i^n - \mu_i^* - an^{-c}\nabla g(\mu_i^n)|^2\right] \leq \left(1 - 2D_1 an^{-c} + a^2 D_2^2 n^{-2c}\right) \mathrm{E}\left[|\mu_i^n - \mu_i^*|^2\right].$$
(3.36)

Then, we have

$$\left(1 + \frac{a}{D_3}n^{-c}\right)\mathrm{E}\left[|\mu_i^n - \mu_i^* - an^{-c}\nabla g(\mu_i^n)|^2\right] \leq$$
$$\left(1 - 2D_1 an^{-c} + \frac{a}{D_3}n^{-c} + \mathcal{O}(n^{-c})\right)\mathrm{E}\left[|\mu_i^n - \mu_i^*|^2\right].$$
(3.37)

For n sufficiently large (e.g., $n > N_0$) and D_3 properly selected in (3.33), there exists $q > D_1 a$ such that

$$\left(1 + \frac{a}{D_3}n^{-c}\right)\mathrm{E}\left[|\mu_i^n - \mu_i^* - an^{-c}\nabla g(\mu_i^n)|^2\right] \leq \left(1 - qn^{-c}\right)\mathrm{E}\left[|\mu_i^n - \mu_i^*|^2\right].$$
(3.38)

Considering (3.38), (3.35) becomes

$$\mathrm{E}\left[|\mu_i^{n+1} - \mu_i^*|^2\right] \leq (1 - qn^{-c})\mathrm{E}[|\mu_i^n - \mu_i^*|^2] + a^2 K_v n^{-2c-b}$$
$$+ a^2 K_e^2 n^{-2c-2b} + aD_3 K_e^2 n^{-c-2b}.$$
(3.39)

Notice that there is a positive real constant (i.e., D_4) such that for $n > N_0$, $a^2 K_v n^{-2c-b} + a^2 K_e^2 n^{-2c-2b} \leq D_4 n^{-2c-b}$. For $n > N_0$, iteration of (3.39) results in

$$\mathrm{E}\left[|\mu_i^{n+1} - \mu_i^*|^2\right] \leq A_{N_0,n}\mathrm{E}\left[|\mu_i^{N_0+1} - \mu_i^*|^2\right] + aD_3 K_e^2 \sum_{m=N_0+1}^{n} A_{mn}m^{-c-2b}$$

$$+ D_4 \sum_{m=N_0+1}^{n} A_{mn}m^{-2c-b}$$
(3.40)

where $A_{mn} = \begin{cases} \prod_{h=m+1}^{n}(1 - qh^{-c}), & 0 \leq m < n \\ 1, & m = n \end{cases}$. The rest of the proof is divided into two steps. First, it is assumed that $0.5 < c < 1$. Then, the case in which $c = 1$ is considered.

Step 1 $(0.5 < c < 1)$

It is well-known that

$$|A_{mn}| \leq e^{-q\sum_{h=m+1}^{n}h^{-c}}$$
$$\leq e^{qm^{-c} - q\int_m^n x^{-c}dx} = e^{qm^{-c} - \frac{q(n^{1-c} - m^{1-c})}{1-c}}.$$
(3.41)

Therefore, the first term of (3.40) is of $\mathcal{O}\left(e^{-n^{1-c}}\right)$. The second term of (3.40) can be written as

$$\sum_{m=N_0+1}^{n} \Lambda_{mn} m^{-c-2b} = n^{-2b} \sum_{m=N_0+1}^{n} A_{mn} m^{-c} \tag{3.42}$$

$$+ \sum_{m=N_0+1}^{n-1} (m^{-2b} - (m+1)^{-2b}) \sum_{r=N_0+1}^{m} A_{rn} r^{-c}.$$

Since $A_{mn} m^{-c} = q^{-1} \left(A_{mn} - A_{m-1,n}\right)$, (3.42) becomes

$$\sum_{m=N_0+1}^{n} A_{mn} m^{-c-2b} = q^{-1} n^{-2b} \left(1 - A_{N_0 n}\right) \tag{3.43}$$

$$+ q^{-1} \sum_{m=N_0+1}^{n-1} (m^{-2b} - (m+1)^{-2b}) \left(A_{mn} - A_{N_0 n}\right).$$

Taking into consideration that $A_{N_0 n} = \mathcal{O}\left(e^{-n^{1-c}}\right)$ and $b > 1 - c$, then $e^{-n^{1-c}} = \mathcal{O}\left(n^{-2b-1}\right)$. Also,

$$\sum_{m=N_0+1}^{n-1} (m^{-2b} - (m+1)^{-2b}) A_{mn} = \sum_{m=N_0+1}^{n-1} (2bm^{-2b-1} + \mathcal{O}(m^{-2b-2})) A_{mn}$$

$$\leq \sum_{m=N_0+1}^{n} D_5 m^{-c-2b} A_{mn} + \mathcal{O}(m^{-2b-1}). \tag{3.44}$$

where D_5 is a positive real constant. Subsequently, (3.43) becomes

$$\sum_{m=N_0+1}^{n} A_{mn} m^{-c-2b} \leq q^{-1} n^{-2b} + D_5 q^{-1} \sum_{m=N_0+1}^{n} A_{mn} m^{-c-2b} + \mathcal{O}(n^{-2b-1}). \tag{3.45}$$

Consequently, we have

$$\sum_{m=N_0+1}^{n} A_{mn} m^{-c-2b} \leq \frac{q^{-1}}{1 - D_5 q^{-1}} n^{-2b} + \mathcal{O}(n^{-2b-1}). \tag{3.46}$$

Similarly,

$$\sum_{m=N_0+1}^{n} A_{mn} m^{-2c-b} \leq \frac{q^{-1}}{1 - D_5 q^{-1}} n^{-b-c} + \mathcal{O}(n^{-b-c-1}). \tag{3.47}$$

As a result,

$$E\left[|\mu_i^{n+1} - \mu_i^*|^2\right] \leq \frac{aD_3 K_e^2 q^{-1}}{1 - D_5 q^{-1}} n^{-2b} + \frac{D_4 q^{-1}}{1 - D_5 q^{-1}} n^{-b-c} + \mathcal{O}\left(n^{-2b-1} + n^{-b-c-1}\right). \tag{3.48}$$

which concludes the proof for this case. ∎
Step 2 (c = 1)
In this case,

$$|A_{mn}| \le e^{-q\sum_{h=m+1}^{n} h^{-1}} \le e^{qm^{-1} - q\int_m^n x^{-1} dx} = e^{qm^{-1}} m^q n^{-q}. \tag{3.49}$$

Therefore, $A_{N_0 n} = \mathcal{O}(n^{-q})$, and hence, the first term of (3.40) is of $\mathcal{O}(n^{-q})$. The second term of (3.40) becomes

$$\sum_{m=N_0+1}^{n} A_{mn} m^{-1-2b} \le n^{-q} e^{qn_0^{-1}} \int_{N_0}^{n} x^{-1-2b+q} dx \le D_6 n^{-2b}. \tag{3.50}$$

where D_6 is a positive real constant. Similarly,

$$\sum_{m=N_0+1}^{n} A_{mn} m^{-2-b} \le D_7 n^{-b-1}. \tag{3.51}$$

where D_7 is a positive real constant. As a consequence, considering that D_8 is a positive real constant,

$$E\left[\left|\mu_i^{n+1} - \mu_i^*\right|^2\right] \le D_8 n^{-q} + D_6 n^{-2b} + D_7 n^{-b-1}$$
$$\le D_8 n^{-D_1 a} + D_6 n^{-2b} + D_7 n^{-b-1}. \tag{3.52}$$

which concludes the proof. ∎

Remark 3.5 *Based on Proposition 3.2, by adjusting the values of b and c, the convergence rate can be properly controlled when $0.5 < c < 1$. If $c = 1$, the magnitude of step size (i.e., a) is another parameter which needs to be designed accurately to achieve the desired convergence rate.*

3.6 Numerical Results

In this section, three numerical examples are presented including one on the performance of the optimal MAC with hopping strategy, the second on the convergence of the learning-based MAC, and the third on the robustness of the optimal MAC under perturbations of PU return probability. In these examples, independent channels are assumed with the same bandwidth $B^i = 1$ and same α_i. We set the same SNR $= \frac{p_k^i}{n_k^i}$ and same channel requirement (i.e., $R_k = 1$) for individual SUs.

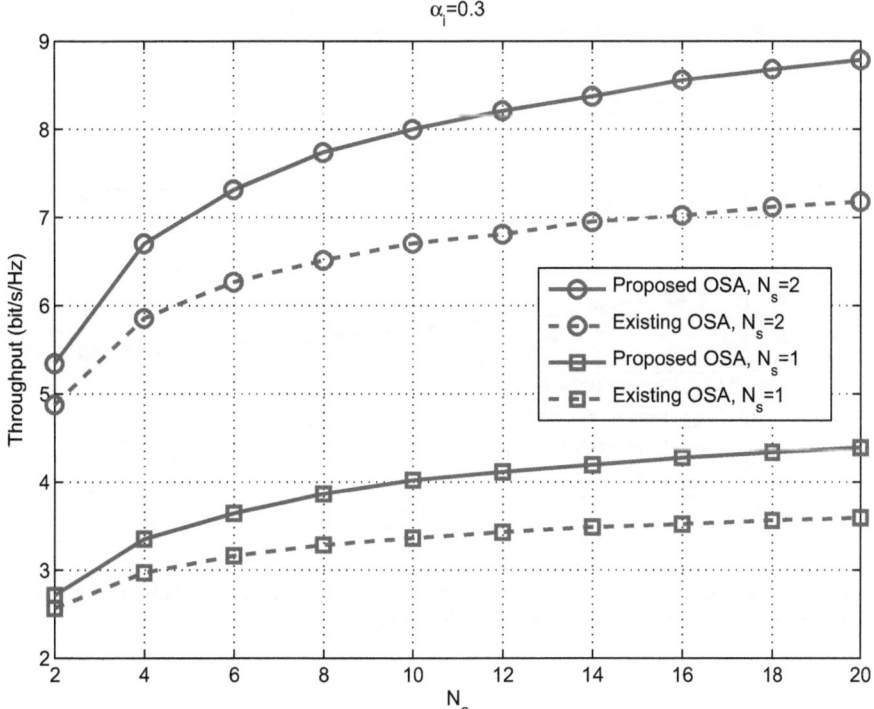

Fig. 3.2 Performance comparison of the proposed MAC and conventional MAC versus the number of idle channels N_a for fixed $\alpha_i = 0.3$ and SNR $= 10$ dB

3.6.1 Performance of the Optimal MAC

First, numerical results are presented to evaluate the performance of the optimal hopping-based MAC. To better understand the proposed approach with $0 \leq \beta_k^i \leq 1$, we compare it with existing MAC strategies (e.g., [1, 9–12, 18, 19]) in which β_k^i is restricted to a binary value. In this example, SU power gains $g_{k,k}^i$ are randomly generated according to the Rayleigh distribution assuming $E[g_{k,k}^i] = 1$.

Figure 3.2 illustrates the overall throughput of all SUs versus number of idle licensed channels (i.e., N_a) for $N_s = 1$ and $N_s = 2$ assuming fixed $\alpha_i = 0.3$ and SNR $= 10$ dB. It is shown that the proposed hopping-based MAC strategy offers a significant system improvement in comparison with the existing MAC protocols. Furthermore, this improvement is an increasing function of N_s. It also demonstrates that the throughput of the SU increases with N_a because distributing the channel requirement into more channels reduces the SU transmission loss due to the PU return and takes advantage of the channel diversity.

On the other hand, with a fixed N_a of 10, Fig. 3.3 shows that the proposed hopping-based MAC strategy offers an overall throughput that is a decreasing func-

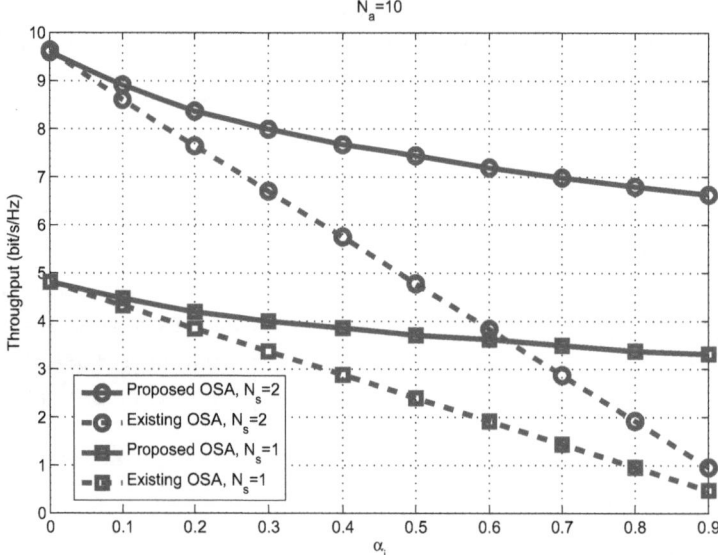

Fig. 3.3 Performance comparison of the proposed MAC and conventional MAC versus the PU return probability α_i for fixed $N_a = 10$ and SNR = 10 dB

tion of α_i with a remarkably lower slope in comparison with the existing MAC protocols. It confirms the advantage of the proposed strategy over the existing MAC protocols, especially at high α_i. In addition, Fig. 3.4 illustrates the overall throughput of all SUs versus SNR of SUs assuming fixed $\alpha_i = 0.3$ and $N_a = 10$. It is shown that throughput improvement increases as the SNR of an individual SU increases.

3.6.2 Convergence of the Learning-Based MAC

Here, the performance of the learning-based MAC protocol is investigated by presenting numerical results which confirm convergence of SUs' activity factors to the optimal values. In this example, it is assumed that $N_s = 3$, $N_a = 3$, $\alpha_i = 0.1$ and SNR = 10 dB. Figures 3.5 and 3.6 illustrate the convergence process of the activity factors of three different SUs in one of the idle licensed channels for two different sets of power gains (i.e., $g_{k,k}^i$). As can be observed, the learning-based scheme takes merely around 100 iterations to quickly converge to optimum solutions. Moreover, it depicts that although activity factors of SUs do not converge by using a fixed learning window, they stay in a close vicinity of the optimal values which is explained by Remark 3.3.

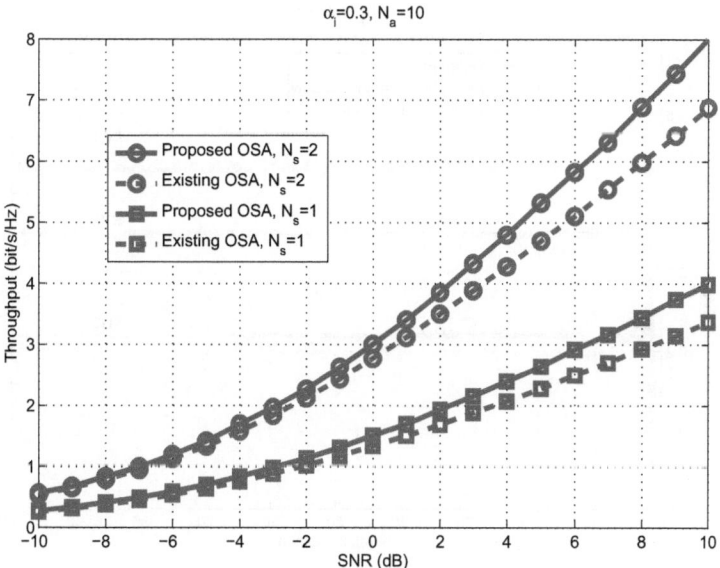

Fig. 3.4 Performance comparison of the proposed MAC and conventional MAC versus SNR for fixed $N_a = 10$ and $\alpha_i = 0.3$

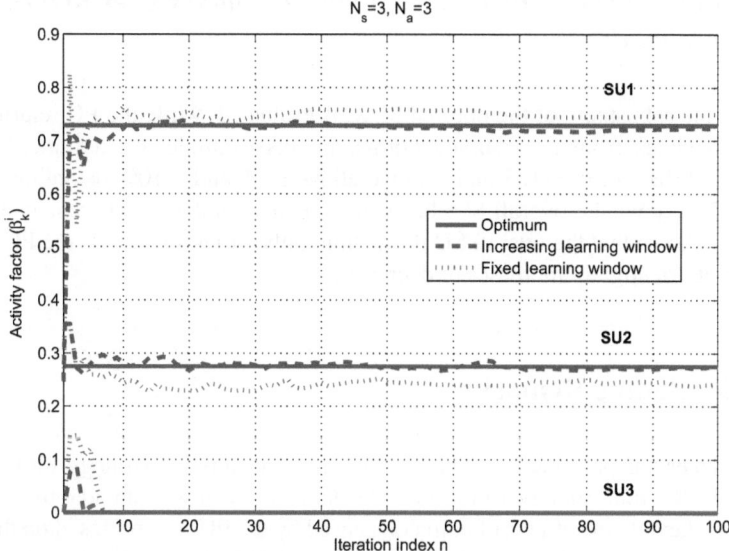

Fig. 3.5 Convergence of the proposed distributed learning-based MAC

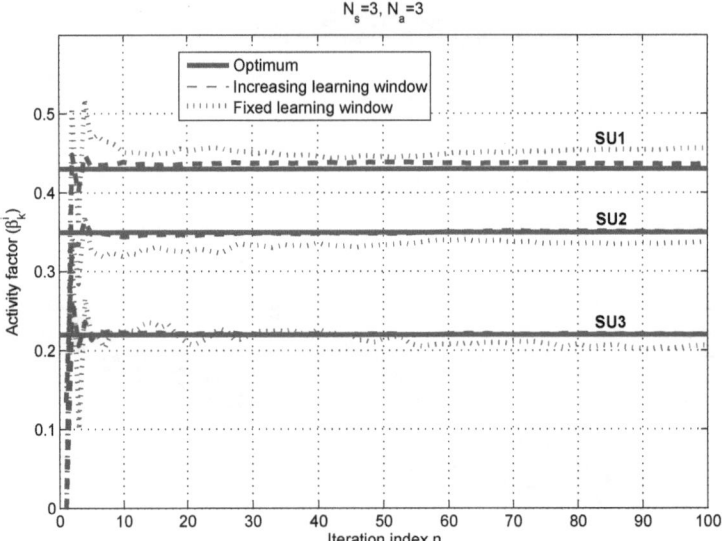

Fig. 3.6 Convergence of the proposed distributed learning-based MAC

3.6.3 Robustness to the Perturbations of Primary User Return Probability

Finally, the robustness of the optimal MAC is evaluated, while the PU return probability deviates from its assumed norms. In this example, we assume $N_s = 2$, SNR = 10 dB and $\alpha_i = 0.3$ and also we allow $\pm 5\%$ and $\pm 10\%$ deviations of α_i. Figure 3.7 shows the overall SU throughput versus number of idle channels (i.e., N_a). It is shown that the overall SU throughput only varies slightly as the PU return probability α_i slightly increases or decreases.

3.7 Concluding Remarks

This chapter has presented an adaptive hopping transmission strategy for OSA, in which the SU transmits over multiple idle channels, each with an adaptive activity factor to alleviate the effects of collision caused by the PU return. Based on the dual decomposition method, a cognitive CSMA-based MAC protocol has been provided, which can be implemented in a distributed manner to determine the optimal activity factors. The optimal values for activity factors reveal the benefits of the proposed approach relative to the existing schemes in which a SU selects one channel to transmit for the entire transmission. Illustrative results confirm performance gains offered by the proposed adaptive hopping access strategy in comparison with the

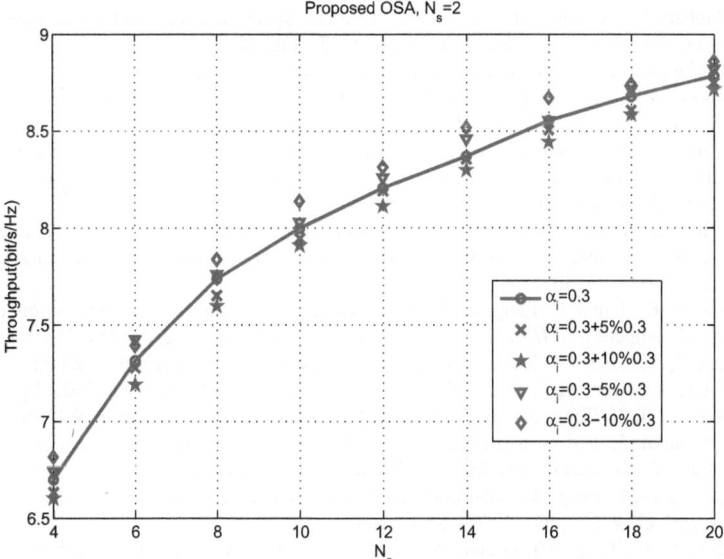

Fig. 3.7 Effect of the PU return probability change on the SU throughput

existing schemes. In addition, a learning-based MAC protocol has been presented, which enables each SU to adapt its activity factors autonomously by learning the other SUs' behavior. Via stochastic gradient search analysis, it has been established that the updated activity factors by SUs converge with probability of 1 to the optimal points. Also, the study on the convergence rate demonstrates that the increasing rate of the learning window size and the decreasing rate of the step size affect how fast the proposed OSA algorithm tracks the optimal values. Illustrative results confirm the validity of the analytical convergence study.

References

1. Q. Zhao, L. Tong, A. Swami, and Y. Chen, "Decentralized cognitive MAC for opportunistic spectrum access in ad hoc networks: A POMDP framework," *IEEE J. Sel. Areas Commun.*, vol. 25, no. 3, pp. 589–600, Apr. 2007.
2. Q. Zhao, S. Geirhofer, L. Tong, and B. M. Sadler, "Opportunistic spectrum access via periodic channel sensing," *IEEE Trans. Signal Process.*, vol. 56, no. 2, pp. 785–796, Feb. 2008.
3. I. F. Akyildiz, W. Y. Lee, M. C. Vuran, and S. Mohanty, "NeXt generation/dynamic spectrum access/cognitive radio wireless networks: A survey," *Computer Networks (Elsevier)*, vol. 50, no. 13, pp. 2127–2159, Sep. 2006.
4. I. F. Akyildiz, W. Y. Lee, M. C. Vuran, and S. Mohanty, "A survey on spectrum management in cognitive radio networks," *IEEE Commun. Mag.*, vol. 46, no. 4, pp. 40–48, Apr. 2008.
5. Q. Zhao and B. M. Sadler, "A survey of dynamic spectrum access: Processing, networking, and regulatory policy," *IEEE Signal Process. Mag.*, vol. 24, no. 3, pp. 79–89, May 2007.

6. B. Jabbari, R. Pickholtz, and M. Norton, "Dynamic spectrum access and management," *IEEE Wireless Commun. Mag.*, vol. 17, no. 4, pp. 6–15, Aug. 2010.
7. L. Berlemann and S. Mangold, *Cognitive Radio and Dynamic Spectrum Access*. Wiley, 2009.
8. E. Hossain, D. Niyato, and Z. Han, *Dynamic Spectrum Access and Management in Cognitive Radio Networks*. Cambridge, 2009.
9. Y. Chen, Q. Zhao, and A. Swami, "Joint design and separation principle for opportunistic spectrum access in the presence of sensing errors," *IEEE Trans. Inf. Theory*, vol. 54, no. 5, pp. 2053–2071, May 2008.
10. Q. Zhao, B. Krishnamachari, and K. Liu, "On myopic sensing for multichannel opportunistic access: Structure, optimality, and performance," *IEEE Trans. Wireless Commun.*, vol. 7, no. 12, pp. 5431–5440, Dec. 2008.
11. L. Lai, H. E. Gamal, H. Jiang, and H. Poor, "Cognitive medium access: Exploration, exploitation and competition," *IEEE Trans. Mobile Comput.*, vol. 10, no. 2, pp. 239–253, Feb. 2011.
12. K. Liu, Q. Zhao, and B. Krishnamachari, "Dynamic multichannel access with imperfect channel state detection," *IEEE Trans. Signal Process.*, vol. 58, no. 5, pp. 2795–2808, May 2010.
13. S. Geirhofer, L. Tong, and B. M. Sadler, "Dynamic spectrum access in WLAN channels: Empirical model and its stochastic analysis," in *Proc. Int. Workshop on Technology and Policy for Accessing Spectrum (TAPAS)*, New York, NY, USA, Aug. 2006.
14. H. Zheng and C. Peng, "Collaboration and fairness in opportunistic spectrum access," in *Proc. IEEE Intl. Conf. Commun. (ICC)*, Seoul, Korea, May 2005.
15. H. Su and X. Zhang, "Cross-layer based opportunistic MAC protocols for QoS provisionings over cognitive radio wireless networks," *IEEE J. Sel. Areas Commun.*, vol. 26, no. 1, pp. 118–129, Jan. 2008.
16. Y. C. Liang, Y. Zeng, E. Peh, and A. T. Hoang, "Sensing-throughput tradeoff for cognitive radio networks," *IEEE Trans. Wireless Commun.*, vol. 7, no. 4, pp. 1326–1337, Apr. 2008.
17. Y. Xu, Y. Sun, Y. Li, Y. Zhao, and H. Zou, "Joint sensing period and transmission time optimization for energy-constrained cognitive radios," *EURASIP J. Wireless Commun. Netw.*, vol. 2010, no. 92, pp. 1–16, Apr. 2010.
18. P. Tehrani, K. Liu, and Q. Zhao, "Opportunistic spectrum access in unslotted primary systems," *Journal of the Franklin Institute (Elsevier)*, vol. 349, no. 3, pp. 985–1010, 2012.
19. S. Shetty, M. Song, C. Xin, and E. K. Park, "A learning-based multiuser opportunistic spectrum access approach in unslotted primary networks," in *Proc. IEEE Intl. Conf. on Computer Commun. (INFOCOM)*, Rio de Janeiro, Brazil, Apr. 2009.
20. D. Palomar and M. Chiang, "A tutorial on decomposition methods for network utility maximization," *IEEE J. Sel. Areas Commun.*, vol. 24, no. 8, pp. 1439–1451, Aug. 2006.
21. P. W. Glynn and P. Heidelberger, "Bias properties of budget constrained simulations," *Operations Research*, vol. 38, no. 5, pp. 801–814, Oct. 1990.
22. H. Kushner and G. Yin, *Stochastic Approximation and Recursive Algorithms and Applications*. New York: Springer, 2003.
23. J. C. Spall, *Introduction to Stochastic Search and Optimization*. Willey, 2003.
24. D. P. Bertsekas and J. N. Tsitsiklis, "Gradient convergence in gradient methods with errors," *SIAM Journal on Optimization*, vol. 10, no. 3, pp. 627–642, Feb. 2000.
25. V. B. Tadić, "Convergence and convergence rate of stochastic gradient search in the case of multiple and non-isolated extrema," in *Proc. IEEE Conf. Decision and Control (CDC)*, Atlanta, GA, USA, Dec. 2010.
26. B. Polyak, *Introduction to Optimization*. New York: Optimization Software, Inc., 1987.
27. J. Sacks, "Asymptotic distribution of stochastic approximation procedures," *The Annals of Mathematical Statistics*, vol. 29, no. 2, pp. 373–405, Jun. 1958.

Chapter 4
Adaptive Carrier Sensing-Based MAC Designs: A Game-Theoretic Approach

4.1 Introduction

In Chap. 3, we have presented an adaptive CSMA scheme to autonomously coordinate spectrum sharing among SUs in the proposed cognitive MAC design. In the proposed CSMA-based MAC protocol, adaptive access probabilities (called activity factors) are adopted and optimized for SUs by taking into account channel qualities, PU return probabilities, and spectrum sharing incentives. Using the proposed adaptive CSMA scheme, in this chapter, we study the SU activity factor allocation problem in a game-theoretic framework. There are three key reasons for using a game-theoretic approach. First, since game theory explicitly recognizes interactions among autonomous SUs, it enables the development of distributed algorithms. Second, game-theoretic algorithms empower us to accelerate the convergence compared to the distributed MAC algorithm proposed in Chap. 3. Third, game theory offers a useful tool to predict, analyze and characterize the long-run behavior of the system, specifically in comparison with the globally optimal solution.

In the proposed game-theoretic algorithm, activity factor selections are made independently and dynamically by each SU which aims to satisfy its own demand despite the imposed sharing incentives. More specifically, we formulate the activity factor optimization problem as an exact potential game and analyze the existence, feasibility, and optimality of the Nash equilibrium (NE). To address incomplete information about the game structure, we study learning approaches, which can respond optimally to the history information and achieve NE points, in terms of information requirements and convergence properties. In light of having perfect information, we establish the convergence of the best-response iterations to a pure NE that is not essentially Pareto-optimal. Aiming to enable equilibrium selection, we introduce the log-linear learning process that assures convergence to the most efficient NE. By introducing noise into the decision making process, the log-linear iterations diverge from the suboptimal NE, while moving in the direction toward the Pareto-optimal NE which is robust to noisy perturbations [1, 2].

© The Author(s) 2014   59
M. Derakhshani, T. Le-Ngoc, *Cognitive MAC Designs for OSA Networks*,
SpringerBriefs in Electrical and Computer Engineering, DOI 10.1007/978-3-319-12649-4_4

Inspired by how the log-linear learning works, we propose a fully-distributed algorithm based on best-response dynamics in which each SU adjusts its activity factors independently by learning from the locally available information. Taking advantage of the noisy observations, we show that the best-response iterations will finally stay in a neighborhood of the Pareto-optimal NE with probability of 1. This can be explained by the fact that the Pareto-optimal NE of the formulated game is the single stochastically stable NE. It is noteworthy that, in comparison with the learning-based algorithm offered in Chap. 3, the proposed game-theoretic algorithm in this chapter appears to have much faster convergence.

Furthermore, in order to address competition among SUs in the adaptive CSMA-based access scheme, we introduce an alternative design objective based on the *achieved* activity factors of SUs instead of *intended* activity factors. Then, the problem of finding optimal activity factors of SUs is cast in a game-theoretic framework to highlight the issues of competition and cooperation among SUs. Subsequently, the existence and characteristics including the uniqueness and efficiency of the NE are investigated. To improve the efficiency of the unique NE in the competitive design, the game is transformed into a more cooperative framework by exploiting a pricing mechanism. Finally, an algorithm based on the best-response dynamics is developed in which each SU independently updates its activity factors until convergence to the unique NE.

The remainder of this chapter is organized as follows. Section 4.2 presents an overview of the system model under consideration. In Sect. 4.3, the OSA MAC design is formulated as an exact potential game. Then, the existence, feasibility and efficiency of the NE for the formulated game are analyzed. Section 4.4 investigates the convergence properties of learning approaches including the best-response dynamics in the presence of perfect information and noisy estimations, and the log-linear dynamics. In Sect. 4.5, the competition among SUs is addressed by introducing an alternative optimization problem in the cognitive MAC design. Finally, Sect. 4.6 presents the concluding remarks.

4.2 System Model

The system model considered in this chapter is similar to the Chap. 3. In particular, we study an OSA network with N_s SUs looking for temporal spectrum availabilities in N_p frequency slots (or channels), licensed to PUs. SUs are assumed to follow a time-slotted transmission aiming to sense the channel before transmission. Each time-slot of equal duration T consists of two periods: sensing of duration τ and transmission of duration $(T - \tau)$. Let $\mathcal{N}_a := \{1, \ldots, N_a\}$ denote the set of N_a channels that are detected idle at the beginning of each time-slot, and hence, can be utilized by N_s SUs.

Regarding the spectrum sharing, it is assumed that SUs share idle channels using the proposed adaptive CSMA approach in Chap. 3, Sect. 3.4. In the proposed CSMA, the SU k enters a competition to access the idle channel $i \in \mathcal{N}_a$ during a transmission slot with a certain probability (called activity factors), $\beta_k^i (0 \leq \beta_k^i \leq 1)$. The activity factors of SUs need to be determined based on the channel qualities, PU return probabilities and sharing incentives. Adaptive activity factors enable prioritizing SUs who gain most from using a channel, and hence, improving channel utilization in comparison with a simple random access scheme.

In a highly dynamic environment such as OSA networks, it is practically essential to find a reasonably good solution which can be obtained in a sufficiently fast manner. To this end, in this work, we study the activity factor optimization problem in (3.8) from a game-theoretic learning perspective which enables distributed implementation and fast convergence to a reasonably good solution.

4.3 Game-Theoretic Design

In this section, we formulate the cognitive MAC design for SUs from a game-theoretic perspective aiming to present a distributed scheme. More specifically, we consider a strategic non-cooperative game in which the players are SUs.

According to the optimization problem in (3.8), each SU could simply maximize its transmission rate (i.e., $\sum_{i=1}^{N_a} \beta_k^i C_k^i (1 - \beta_k^i \alpha_i)$). However, SUs cannot select activity factors which violate the coupling constraints in (3.8b). Since it is difficult for SUs to identify feasible activity factors in advance, we construct an alternative payoff function for SU k as

$$u_k = \sum_{i=1}^{N_a} \beta_k^i C_k^i \left(1 - \beta_k^i \alpha_i\right) - \sum_{i=1}^{N_a} \mu_i \Theta \left(\sum_{j=1}^{N_s} \beta_j^i - 1 \right) \tag{4.1}$$

where $\Theta(x) = \begin{cases} x, & x \geq 0 \\ 0, & x < 0 \end{cases}$ and μ_i's are positive scalars. The second term of (4.1) represents the coupling constraints in (3.8b) by severely punishing the SU who violates each of them.

Let $\boldsymbol{\beta}_k = \left[\beta_k^1, \dots, \beta_k^{N_a} \right]$ be the strategy at SU k and $\boldsymbol{\beta}_{-k}$ be the strategy of all SUs excluding the SU k. Furthermore, the admissible strategies of SU k is defined as

$$\mathscr{B}_k = \left\{ \boldsymbol{\beta}_k : \beta_k^i \in \{0, \frac{1}{S}, \frac{2}{S}, \dots, 1\}, \forall i \in \mathscr{N}_a, \sum_{i=1}^{N_a} \beta_k^i = R_k \right\}. \tag{4.2}$$

Note that β_k^i takes discrete values in the proposed CSMA-based algorithm. Then, we can define the non-cooperative game for the spectrum access design of SUs as

$$\mathscr{G} = [\mathscr{N}_s, \{\mathscr{B}_k\}_{k \in \mathscr{N}_s}, \{u_k\}_{k \in \mathscr{N}_s}] \tag{4.3}$$

where $\mathcal{N}_s = \{1,\ldots,N_s\}$ is the set of players of the game (i.e., SUs), $\mathcal{B}_k$ is the activity factor strategy set of the SU k, and u_k is the corresponding payoff function of the SU k defined on the set of pure-strategy profiles $\mathcal{B} = \mathcal{B}_1 \times \ldots \times \mathcal{B}_{N_s}$.

An exact potential game is a strategic game in which the incentive of all players to change their strategies can be expressed in a global potential function. The potential games are easy to analyze since improving each player's utility also increases the value of a potential function [3]. In the following theorem, we demonstrate that the game $\mathcal{G}$ falls into the framework of exact potential games.

Theorem 4.1 $\mathcal{G}$ *is an exact potential game with the potential function,*

$$\Phi = \sum_{j=1}^{N_s}\sum_{i=1}^{N_a} \beta_j^i C_j^i (1 - \beta_j^i \alpha_i) - \sum_{i=1}^{N_a} \mu_i \Theta\left(\sum_{j=1}^{N_s}\beta_j^i - 1\right). \tag{4.4}$$

Proof: It is clear that the game $\mathcal{G}$ satisfies the exact potential game definition [3],

$$u_k(\boldsymbol{\beta}_k, \boldsymbol{\beta}_{-k}) - u_k(\boldsymbol{\beta}_k', \boldsymbol{\beta}_{-k}) = \Phi(\boldsymbol{\beta}_k, \boldsymbol{\beta}_{-k}) - \Phi(\boldsymbol{\beta}_k', \boldsymbol{\beta}_{-k}), \forall \boldsymbol{\beta}_k, \boldsymbol{\beta}_k' \in \mathcal{B}_k, \forall k \in \mathcal{N}_s. \tag{4.5}$$

Thus, $\mathcal{G}$ is an exact potential game and Φ is the potential function of $\mathcal{G}$. ∎

Conceptually, a strategic game can reach a steady-state NE point, if it exists, from which no player can improve its utility by changing its own strategy unilaterally [4]. In other words, a strategy profile $\boldsymbol{\beta}^* = \{\boldsymbol{\beta}_k^*\}_{k=1}^{N_s} \in \mathcal{B}$ is a NE if and only if

$$u_k(\boldsymbol{\beta}_k^*, \boldsymbol{\beta}_{-k}^*) \geq u_k(\boldsymbol{\beta}_k', \boldsymbol{\beta}_{-k}^*), \forall \boldsymbol{\beta}_k' \in \mathcal{B}_k, \forall k \in \mathcal{N}_s. \tag{4.6}$$

We are interested to investigate the existence and characteristics including the feasibility and efficiency of NE for the game $\mathcal{G}$.

4.3.1 Existence of the Nash Equilibrium

First of all, we study the existence of NE of the game $\mathcal{G}$ in the following theorem based on the properties of the potential games.

Theorem 4.2 *The game $\mathcal{G}$ admits at least one pure-strategy NE.*

Proof: This theorem comes directly from Corollary 4 in [5], which states every finite potential game $\mathcal{G}$ has at least one pure-strategy NE. ∎

Remark 4.1 *In general, the pure-strategy NE of game $\mathcal{G}$ may not be unique.*

4.3.2 Feasibility of the Nash Equilibrium

Since the optimization problem in (3.8) has the coupling constraints in (3.8b) which are merged in the payoff functions in the formulated game $\mathscr{G}$, it is required to verify if an arbitrary pure-strategy NE is feasible, i.e., satisfying the constraints $\sum_{k=1}^{N_s} \beta_k^i \leq 1, i = 1, \ldots, N_a$. Thus, the following theorem presents conditions that assure the feasibility of pure-strategy NEs.

Theorem 4.3 *All pure-strategy NEs of the game $\mathscr{G}$ must be feasible if*

$$\mu_i > \mu_{th}, \forall i \in \mathcal{N}_a \tag{4.7}$$

where

$$\mu_{th} = \max_{k \in \mathcal{N}_s, \boldsymbol{\beta}_k \in \mathscr{B}_k} \left(s \sum_{i=1}^{N_a} \beta_k^i C_k^i \left(1 - \beta_k^i \alpha_i\right) \right). \tag{4.8}$$

Proof: Suppose that $\boldsymbol{\beta} = \{\boldsymbol{\beta}_k\}_{k=1}^{N_s}$, where $\boldsymbol{\beta}_k \in \mathscr{B}_k$, is a pure-strategy NE of game $\mathscr{G}$, but it is not feasible, i.e., it violates at least one of the constraints in (3.8b). Further, suppose that m is the index of channel which has the most severe violation, i.e., $m = \arg\max_{i \in \mathcal{N}_a} \left(\sum_{j=1}^{N_s} \beta_j^i - 1 \right)$. Thus, $\sum_{j=1}^{N_s} \beta_j^m > 1$. Considering $k \in \mathcal{N}_s$ such that $\beta_k^m > 0$, from (4.1), we have

$$u_k(\boldsymbol{\beta}_k, \boldsymbol{\beta}_{-k}) = \sum_{i=1}^{N_a} \beta_k^i C_k^i \left(1 - \beta_k^i \alpha_i\right) - \mu_m \left(\sum_{j=1}^{N_s} \beta_j^m - 1 \right) - \sum_{i=1, i \neq m}^{N_a} \mu_i \Theta \left(\sum_{j=1}^{N_s} \beta_j^i - 1 \right). \tag{4.9}$$

Assuming that $\sum_{j=1}^{N_s} \beta_j^m = 1 + \varepsilon$ where ε is a positive discrete value (i.e., $\varepsilon \geq \frac{1}{s}$), there exists a channel n such that $\sum_{j=1}^{N_s} \beta_j^n \leq 1 - \xi$ where $\frac{1}{s} \leq \xi < \min(1, \varepsilon)$ since $N_a \geq N_s$. Then, an alternative admissible strategy $\tilde{\boldsymbol{\beta}}_k = \left[\tilde{\beta}_k^i \right]_{i \in \mathcal{N}_a} \in \mathscr{B}_k$ can be con-

structed such that $\tilde{\beta}_k^i = \begin{cases} \beta_k^i, & i \neq m, n \\ \beta_k^i - \delta, & i = m \\ \beta_k^i + \delta, & i = n \end{cases}$ where δ is a positive discrete value

(i.e., $\delta \in \{\frac{1}{s}, \frac{2}{s}, \ldots, 1\}$) satisfying $\delta \leq \xi$, $0 \leq \beta_k^m - \delta \leq 1$ and $0 \leq \beta_k^n + \delta \leq 1$. Note that $\sum_{j=1}^{N_s} \tilde{\beta}_j^n - 1 = \delta - \xi \leq 0$ since $\delta \leq \xi$, and hence, $\Theta \left(\sum_{j=1}^{N_s} \tilde{\beta}_j^n - 1 \right) = \Theta \left(\sum_{j=1}^{N_s} \beta_j^n - 1 \right) = 0$. Then, $u_k \left(\tilde{\boldsymbol{\beta}}_k, \boldsymbol{\beta}_{-k} \right)$ can be obtained as

$$u_k(\tilde{\boldsymbol{\beta}}_k, \boldsymbol{\beta}_{-k}) = \sum_{i=1}^{N_a} \tilde{\beta}_k^i C_k^i (1 - \tilde{\beta}_k^i \alpha_i) - \sum_{i=1}^{N_a} \mu_i \Theta \left(\sum_{j=1}^{N_s} \tilde{\beta}_j^i - 1 \right)$$

$$= \sum_{i=1}^{N_a} \tilde{\beta}_k^i C_k^i (1 - \tilde{\beta}_k^i \alpha_i) - \mu_m \Theta \left(\sum_{j=1}^{N_s} \beta_j^m - 1 - \delta \right) - \sum_{i=1, i \neq m}^{N_a} \mu_i \Theta \left(\sum_{j=1}^{N_s} \beta_j^i - 1 \right).$$

$$(4.10)$$

Thus, from (4.9) and (4.10),

$$u_k(\boldsymbol{\beta}_k, \boldsymbol{\beta}_{-k}) - u_k(\tilde{\boldsymbol{\beta}}_k, \boldsymbol{\beta}_{-k}) = \sum_{i=1}^{N_a} \beta_k^i C_k^i \left(1 - \beta_k^i \alpha_i \right) - \sum_{i=1}^{N_a} \tilde{\beta}_k^i C_k^i (1 - \tilde{\beta}_k^i \alpha_i)$$

$$- \mu_m \left(\sum_{j=1}^{N_s} \beta_j^m - 1 \right) + \mu_m \Theta \left(\sum_{j=1}^{N_s} \beta_j^m - 1 - \delta \right).$$

$$(4.11)$$

Since $\delta \leq \varepsilon$, we have

$$\mu_m \Theta \left(\sum_{j=1}^{N_s} \beta_j^m - 1 - \delta \right) - \mu_m \left(\sum_{j=1}^{N_s} \beta_j^m - 1 \right) = \mu_m \Theta(\varepsilon - \delta) - \mu_m \varepsilon \leq -\mu_m \delta.$$

$$(4.12)$$

Subsequently, from (4.11) and (4.12),

$$u_k(\boldsymbol{\beta}_k, \boldsymbol{\beta}_{-k}) - u_k(\tilde{\boldsymbol{\beta}}_k, \boldsymbol{\beta}_{-k}) \leq \sum_{i=1}^{N_a} \beta_k^i C_k^i \left(1 - \beta_k^i \alpha_i \right) - \sum_{i=1}^{N_a} \tilde{\beta}_k^i C_k^i (1 - \tilde{\beta}_k^i \alpha_i) - \mu_m \delta.$$

$$(4.13)$$

Under the assumption of $\mu_i > \mu_{th}, \forall i \in \mathcal{N}_a$, from (4.8), we have

$$\mu_m > S \sum_{i=1}^{N_a} \beta_k^i C_k^i \left(1 - \beta_k^i \alpha_i \right). \tag{4.14}$$

Since $\delta \geq \frac{1}{S}$, from (4.14), it can be concluded that

$$\sum_{i=1}^{N_a} \beta_k^i C_k^i \left(1 - \beta_k^i \alpha_i \right) - \mu_m \delta < 0. \tag{4.15}$$

Thus, based on (4.13) and (4.15),

$$u_k(\boldsymbol{\beta}_k, \boldsymbol{\beta}_{-k}) - u_k(\tilde{\boldsymbol{\beta}}_k, \boldsymbol{\beta}_{-k}) \leq 0. \tag{4.16}$$

Note that this contradicts the assumption that $\boldsymbol{\beta} = \{\boldsymbol{\beta}_k\}_{k=1}^{N_s}$ is a pure-strategy NE of game $\mathcal{G}$ according to the definition of NE in (4.6). Thus, $\boldsymbol{\beta}$ is not a pure-strategy

NE of $\mathscr{G}$. Hence, it can be concluded that all pure-strategy NEs must be feasible if $\mu_i > \mu_{th}, \forall i \in \mathscr{N}_a$. ∎

Theorem 4.3 ensures that, by properly designing μ_i's, the payoff functions in (4.1) can guarantee the feasibility of the steady states of the system.

4.3.3 Efficiency of the Nash Equilibrium

The other aspect that we study is how efficient the NE of game $\mathscr{G}$ is in comparison with the optimal solution of (3.8). The following theorem specifies the relationship between the optimal solution and the NE of the game $\mathscr{G}$.

Theorem 4.4 *The optimal solution of (3.8) is the Pareto-optimal pure-strategy NE of $\mathscr{G}$ if $\mu_i > \mu_{th}, \forall i \in \mathscr{N}_a$.*

Proof: Assume that $\boldsymbol{\beta} = \{\boldsymbol{\beta}_k\}_{k=1}^{N_s}$ is the optimal solution of (3.8). Assuming $\mu_i > \mu_{th}, \forall i \in \mathscr{N}_a$, $\boldsymbol{\beta}$ is the maximizer of the potential function Φ. Based on Theorem 2 in [3], the maximizer of the potential function is the NE of the potential game. Hence, $\boldsymbol{\beta}$ is the NE of the game $\mathscr{G}$.

Subsequently, we need to establish that $\boldsymbol{\beta}$ is the Pareto-optimal NE. Assume that $\boldsymbol{\beta}$ is not Pareto-optimal. Then, there exists an arbitrary strategy profile $\boldsymbol{\beta}' = \{\boldsymbol{\beta}'_k\}_{k=1}^{N_s}$ such that

$$u_k(\boldsymbol{\beta}'_k, \boldsymbol{\beta}'_{-k}) \geq u_k(\boldsymbol{\beta}_k, \boldsymbol{\beta}_{-k}), \forall k \in \mathscr{N}_s, k \neq j \tag{4.17}$$

and, for some j,

$$u_j(\boldsymbol{\beta}'_j, \boldsymbol{\beta}'_{-j}) > u_j(\boldsymbol{\beta}_j, \boldsymbol{\beta}_{-j}). \tag{4.18}$$

As a result,

$$\sum_{k=1}^{N_s} u_k(\boldsymbol{\beta}') > \sum_{k=1}^{N_s} u_k(\boldsymbol{\beta}). \tag{4.19}$$

Since $\boldsymbol{\beta}$ is a NE of the game $\mathscr{G}$, it is feasible based on Theorem 4.3 (i.e., $\sum_{i=1}^{N_a} \mu_i \Theta\left(\sum_{j=1}^{N_s} \beta_j^i - 1\right) = 0$). Then, we have

$$\Phi(\boldsymbol{\beta}) = \sum_{k=1}^{N_s} u_k(\boldsymbol{\beta}). \tag{4.20}$$

Furthermore, considering that $\mu_i, \forall i \in \mathscr{N}_a$ are positive scalars, from (4.1) and (4.4), for an arbitrary $\boldsymbol{\beta}'$, we have

$$\Phi(\boldsymbol{\beta}') \geq \sum_{k=1}^{N_s} u_k(\boldsymbol{\beta}'). \tag{4.21}$$

Consequently, based on (4.19), (4.20) and (4.21),

$$\Phi(\boldsymbol{\beta}') > \Phi(\boldsymbol{\beta}). \tag{4.22}$$

This contradicts the fact that $\boldsymbol{\beta}$ is the maximizer of the potential function Φ. Thus, the optimal solution of (3.8) is the Pareto-optimal pure-strategy NE of the game $\mathscr{G}$. ∎

Remark 4.2 *In the next section, learning-based iterative algorithms are proposed that enable the convergence to the Pareto-optimal pure-strategy NE.*

4.4 Iterative Learning-based MAC with Perfect and Noisy Observations

Assuming that the rationality of players and the structure of the game are common knowledge, equilibrium can be observed as a result of analysis and introspection of the players. Otherwise, under assumption of bounded rationality or partial information, equilibrium may arise as a consequence of a long-run learning process [6]. In this section, aiming to achieve an equilibrium of the game $\mathscr{G}$, we discuss learning approaches in terms of information requirements and convergence properties.

4.4.1 Best-Response Dynamics with Perfect Observations

To reach an equilibrium of the game $\mathscr{G}$, first, we present a simple learning algorithm for activity factor selection, based on the asynchronous best-response dynamics [6]. In particular, at each time $t \in \{0,1,2,\ldots\}$, exactly one SU $k \in \mathcal{N}_s$, is randomly selected to revise its activity factors based on the best-response dynamics defined as

$$\boldsymbol{\beta}_k[t] = \arg \max_{\boldsymbol{\beta}'_k \in \mathscr{B}_k} u_k(\boldsymbol{\beta}'_k, \boldsymbol{\beta}_{-k}[t-1]). \tag{4.23}$$

Under the perfect knowledge of current strategies of the other SUs (i.e., $\boldsymbol{\beta}_{-k}[t-1]$), the convergence of the proposed game-theoretic algorithm is established in the following theorem.

Theorem 4.5 *The learning algorithm under asynchronous best-response dynamics converges with probability of 1 to a pure-strategy NE of the game $\mathscr{G}$ from any initial strategy point.*

Proof: Based on Theorem 19 in [4], in a finite exact potential game, best-response dynamics will converge with probability of 1 to a pure-strategy NE in finite steps. Accordingly, in the game $\mathscr{G}$, the best-response iterations will converge to a pure-strategy NE. ∎

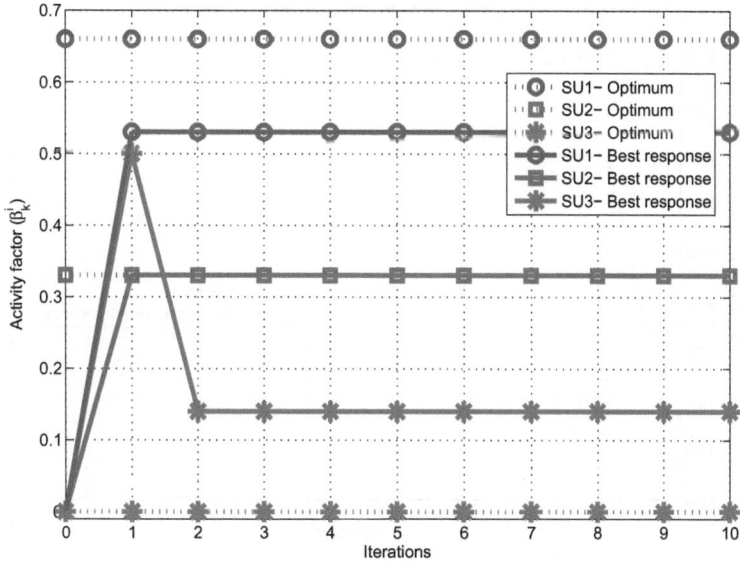

Fig. 4.1 Convergence of the SU activity factors with best-response dynamics

To verify convergence of the best-response dynamics, we provide a numerical result. In this example, we assume independent channels with the same bandwidth $B^i = 1$ and the same α_i. We set the same SNR $= \frac{p_k^i}{n_k^i} = 10$ dB and $R_k = 1$ for individual SUs. Furthermore, we assume $N_s = 3$, $N_a = 3$ and $\alpha_i = 0.1$. Note that we use the same example for all simulation results in this chapter.

Figure 4.1 demonstrates the convergence process of the activity factors of three SUs in the first idle licensed channel for a certain channel realization. All SUs start by setting their initial values of their activity factors to zero. In each iteration, they sequentially play to optimize their own payoff functions, and update their activity factors based on the best-response dynamics in (4.23) (e.g., in the following order: SU3, SU2, SU1). Note that each SU updates its activity factors in three different channels simultaneously. However, the simulation results for only channel 1 are demonstrated.

In iteration 1, SU3 plays followed by SU2 and SU1 to obtain $\beta_3^1[1] = 0.5$, $\beta_2^1[1] = 0.33$, and $\beta_1^1[1] = 0.53$, which result in $\sum_{k=1}^{3} \beta_k^1 \geq 1$. In iteration 2, SU3 is penalized for this excessive amount, and, for its specific channel realization, Fig. 4.1 indicates that the SU3 is forced to reduce its activity factor to $\beta_3^1[2] = 0.14$ in order to keep $\sum_{k=1}^{3} \beta_k^1 \leq 1$. Subsequently, SU2 and SU1 do not need to change their strategies, and maintain $\beta_2^1[2] = 0.33$, and $\beta_1^1[2] = 0.53$. From iteration 3, since $\sum_{k=1}^{3} \beta_k^1 \leq 1, i = 1, \ldots, 3$, SUs actually optimize their own throughput, i.e., their own payoff functions become $u_k = \sum_{i=1}^{3} \beta_k^i C_k^i \left(1 - \beta_k^i \alpha_i\right)$. Thus, their strategies do not change any more. Note that, for different channel realizations, it is possible that all SUs change their

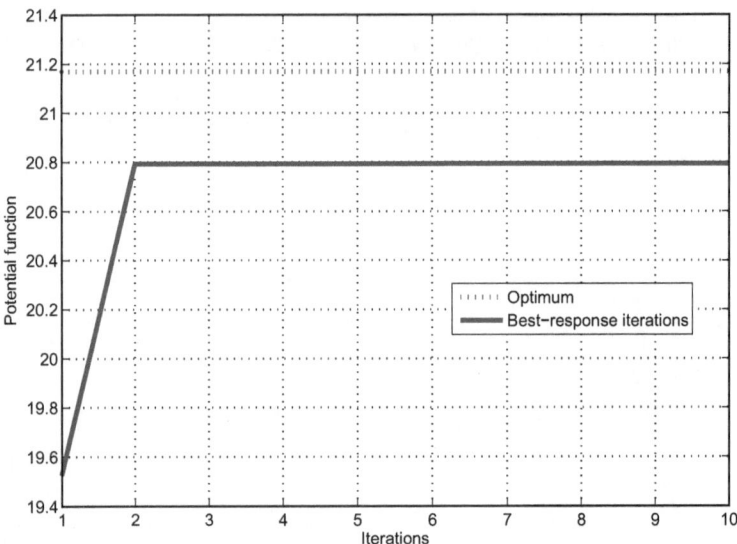

Fig. 4.2 Convergence of the potential function with best-response dynamics

activity factors to satisfy $\sum_{k=1}^{N_s} \beta_k^i \leq 1$ in iteration 2. Furthermore, the process of reducing $\sum_{k=1}^{N_s} \beta_k^i$ may take more than two iterations.

Additionally, Fig. 4.2 shows the convergence of the potential function. As evident from Figs. 4.1 and 4.2, in less than 10 iterations, the activity factor selection algorithm converges to a NE which is not essentially Pareto-optimal. Note that each iteration corresponds to a complete update by all the SUs.

According to Theorem 5, the best-response iterations will converge to a pure-strategy NE which is not necessarily the maximizer of the potential function Φ (i.e., the globally optimal solution of (3.8)). Since NE could be highly inefficient with regards to the network-level objective, it is thus crucial to find a learning process to reach the socially optimal solution.

4.4.2 Log-Linear Dynamics with Perfect Observations

Since the basic best-response dynamics suffers from multiple rest points (e.g., any pure-strategy NE), it is essential to introduce a learning process that can select an appropriate equilibrium. Aiming to enable equilibrium selection, log-linear learning has been proposed as a perturbed best-response process which guarantees convergence to the most efficient NE for potential games [1, 2]. The basic idea behind the equilibrium selection in the log-linear learning is to introduce noise into decision making process which enables categorizing equilibria based on their stability characteristics. This noise allows players to select suboptimal actions with a certain probability which is attached with the magnitude of the payoff difference of the best response and the suboptimal action.

In the log-linear learning, SUs are assumed to be myopic and boundedly rational. At each time $t > 0$, exactly one SU $k \in \mathcal{N}_s$ is randomly selected to update its action, using a probability distribution over its strategy set in response to the current strategy profile. Let k be the player chosen at time t to revise its action. Then, the SU k will choose action $\boldsymbol{\beta}_k$ given the current strategy profile $\boldsymbol{\beta}_{-k}[t-1]$ with a probability based on the log-linear choice rule

$$\pi_k^{\boldsymbol{\beta}_k}[t] = \frac{e^{\frac{1}{\varepsilon}u_k(\boldsymbol{\beta}_k,\boldsymbol{\beta}_{-k}[t-1])}}{\sum_{\boldsymbol{\beta}'_k \in \mathscr{B}_k} e^{\frac{1}{\varepsilon}u_k(\boldsymbol{\beta}'_k,\boldsymbol{\beta}_{-k}[t-1])}} \tag{4.24}$$

where $0 < \frac{1}{\varepsilon} < \infty$. The scalar $\frac{1}{\varepsilon}$ can be interpreted as the level of rationality of the SUs. In other words, ε shows the level of noise in the SUs' decisions and determines how often SUs choose their best responses. The described rule is called log-linear since the log-likelihood ratio of selecting between two actions is linearly proportional to the difference of corresponding payoffs, given other SUs' actions [7]. As $\varepsilon \to 0$, the log-linear rule approaches to the best-response rule. However, the SU k will choose any action $\boldsymbol{\beta}_k \in \mathscr{B}_k$ with equal probability as $\varepsilon \to \infty$. Thus, for any $0 < \varepsilon < \infty$, SUs explore non-best responses with non-zero probabilities which are exponentially smaller for actions yielding smaller payoffs [1, 2].

Our goal is to characterize the long-run behavior of asynchronous log-linear learning process for game $\mathscr{G}$. To this end, the log-linear dynamic adjustment process is represented as an irreducible and aperiodic Markov chain $\{X_t^\varepsilon\}_{t \in \mathbb{N}}$ on the set of strategy profiles of the game [1, 2]. Subsequently, the stationary distribution, i.e., limiting distribution in a Markov chain, is studied to explain the long-run behavior of this update rule. In [1], the stationary distribution of asynchronous log-linear learning process in game $\mathscr{G}$ is presented as

$$\mathscr{P}^\varepsilon(\boldsymbol{\beta}) = \frac{e^{\frac{1}{\varepsilon}\Phi(\boldsymbol{\beta})}}{\sum_{\boldsymbol{\beta} \in \mathscr{B}} e^{\frac{1}{\varepsilon}\Phi(\boldsymbol{\beta})}}. \tag{4.25}$$

According to (4.25), $\mathscr{P}^\varepsilon(\boldsymbol{\beta})$ (i.e., the probability that $X_t^\varepsilon = \boldsymbol{\beta}$ for sufficiently large times $t > 0$) can be expressed as an explicit function of the potential function. A strategy profile $\boldsymbol{\beta} = \{\boldsymbol{\beta}_k\}_{k=1}^{N_s}$ is said to be stochastically stable if $\lim_{\varepsilon \to 0} \mathscr{P}^\varepsilon(\boldsymbol{\beta}) > 0$. Consequently, the stochastically stable strategy can be computed based on (4.25).

Corollary 4.1 *In game $\mathscr{G}$, the only stochastically stable strategy profile of asynchronous log-linear learning process is the maximizer of potential function Φ which is the NE of the potential game $\mathscr{G}$ based on Theorem 2 in [3]. Assuming $\mu_i > \mu_{th}, \forall i \in \mathcal{N}_a$, the optimal solution of (3.8) is equal to the maximizer of the potential function. As a result, the optimal solution of (3.8) is the only stochastically stable NE of the game $\mathscr{G}$ assuming $\mu_i > \mu_{th}, \forall i \in \mathcal{N}_a$.*

The importance of this Corollary–which comes directly from Corollary 1 in [1] for exact potential games–is two-fold, i.e., convergence and equilibrium selection. In particular, it explains that the asynchronous log-linear learning guarantees convergence to a set of Nash equilibria and more specifically enables equilibrium refinement. In other words, the log-linear learning assures convergence to the potential maximizer which is equal to the globally optimal solution in the game $\mathscr{G}$ assuming $\mu_i > \mu_{th}, \forall i \in \mathscr{N}_a$.

To verify the convergence properties of log-linear learning, Figs. 4.3 and 4.4 show the convergence process of the activity factors of three SUs in the first idle licensed channel for a certain channel realization and the convergence of the potential function, respectively. They confirm that the log-linear iterations lead to the globally optimal solution despite the best-response dynamics. As can be observed, it takes merely around 20 iterations to quickly converge to the optimum solution.

The other aspect which affects the practicality of a learning process is the convergence speed. From Proposition 152 in [8], assuming $\mu_i > \mu_{th}$, the convergence time of the log-linear learning to be η-close to the optimal solution is in the order of

$$N_s \log \log(N_s) + \log\left(\frac{1}{\eta}\right) \tag{4.26}$$

for any initial condition if the rationality level (i.e., $\frac{1}{\varepsilon}$) is sufficiently large. According to (4.26), the convergence time is linearly proportional to the number of SUs using the log-linear learning in game $\mathscr{G}$.

Up to this point, we study the best-response dynamics and the log-linear dynamics assuming the perfect knowledge of the sum of activity factors of all N_s SUs in an idle channel, i.e., $\sum_{k=1}^{N_s} \beta_k^i$. Such information can be obtained with the aid of a central coordinator or heavy exchange of overhead information, which causes high complexity and results in an un-scalable system. It is thus crucial that SUs learn this information to adjust their activity factors.

4.4.3 Best-Response Dynamics with Noisy Observations

In Chap. 3, Sect. 3.4, we have studied how to use the capturing status feedbacks of the proposed adaptive CSMA scheme, $y_k^i(t_j)$, to estimate the sum of activity factors of all SUs in each channel. It is shown that $\beta^i = \sum_{k=1}^{N_s} \beta_k^i$ can be updated after each window of S' sub-slots as $\hat{\beta}^i \simeq \beta_k^i + \left(\frac{S' \cdot \beta_k^i}{\sum_{l=f+1}^{f+S'} y_k^i(t_l)}\right) - 1$. Since estimation with a limited number of samples suffers from random errors, it is shown that $\hat{\beta}^i = \beta^i + w$ where $E[w]$ and $var[w]$ are of $\mathscr{O}\left((S')^{-1}\right)$.

From (4.1), it is clear that the estimation noise of β^i will cause a bias b ($|b| \leq |w|$) in u_k, and hence, best-response iterations in (4.23) will also involve random errors.

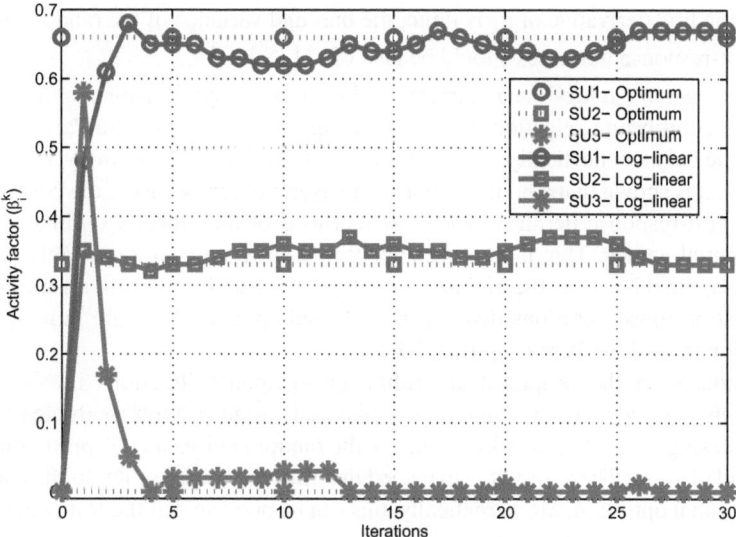

Fig. 4.3 Convergence of the SU activity factors with log-linear dynamics

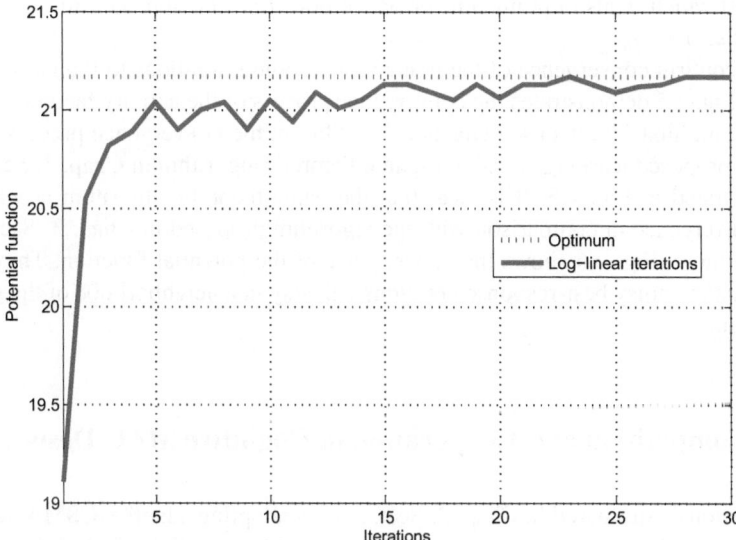

Fig. 4.4 Convergence of the potential function with log-linear dynamics

Since the first derivative of u_k is finite, the bias and variance of the random noise in the best-response iterations should be also of $\mathscr{O}\left((S')^{-1}\right)$.

As shown in the log-linear learning, adding noise to the decision making process enables equilibrium selection, taking advantage of the fact that the Pareto-optimal NE is the only stochastically stable NE in potential games. Accordingly, in Theorem 3 of [9], it is shown that a bounded noise will asymptotically ensure the convergence of the best-response iterations to a neighborhood of the globally optimal solution in potential games. That is because suboptimal NE points are less stable than the Pareto-optimal NE (i.e., the global optimum) in a sense that a small noise can cause the best-response iterations diverge from the suboptimal NE while moving in the direction toward the Pareto-optimal NE.

Similarly, in the proposed algorithm, best-response iterations involve errors although they are random with bounded bias and variance. With a sufficiently large or increasing estimation window (i.e., S'), the random noise can be approximated as a bounded noise. Therefore, it is expected that the best-response iterations converge to the global optimum. Mathematically, this can be presented as the following claim.

Claim 4.1 $\forall \xi > 0$, *an estimation window size can be selected (i.e., $\exists S' > 0$) such that*
$\lim_{t \to \infty} \inf \Phi(\boldsymbol{\beta}[t]) \geq \Phi_{\max} - \xi$ *with probability of 1.*

This claim declares that, by properly designing an estimation window size (i.e., S'), the best-response iterations can get arbitrarily close to the globally optimal solution of (3.8) which is also the maximizer of the potential function (i.e., Φ), assuming $\mu_i > \mu_{th}, \forall i \in \mathcal{N}_a$.

To confirm convergence of the noisy best-response iterations to the global optimum, Fig. 4.5 demonstrates the convergence process of the activity factors of three SUs in the first idle licensed channel. In addition, the convergence process based on the proposed learning-based non-game-theoretic algorithm in Chap. 3, Sect. 3.4 is illustrated in Fig. 4.5. It is clear that the game-theoretic algorithm accelerates the convergence in comparison with the algorithm proposed in Chap. 3, Sect. 3.4. Furthermore, Fig. 4.6 shows the convergence of the potential function. They confirm that the noisy best-response iterations will stay in a neighborhood of the global optimum.

4.5 Competition and Cooperation in Cognitive MAC Design

As previously discussed in Chap. 3, Sect. 3.4, by adopting adaptive CSMA scheme as a decentralized mechanism among SUs, the *achieved* activity factor of each SU depends on the *intended* activity factors of the other SUs, i.e., $\bar{\beta}_k^i \simeq \frac{\beta_k^i}{1 + \sum_{\bar{k}=1, \bar{k} \neq k}^{N_S} \beta_{\bar{k}}^i}$. This is due to the congestion nature of channel contention in carrier sensing. Accordingly, we consider an alternative utility function based on *achieved* activity factor of

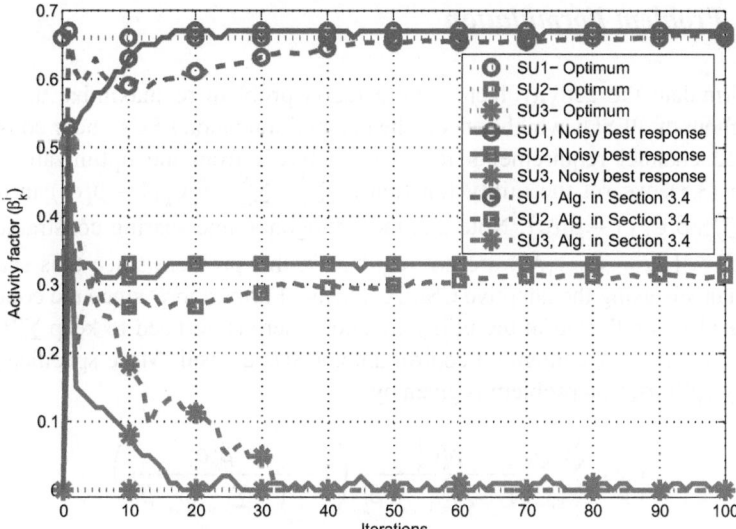

Fig. 4.5 Convergence of the SU activity factors with noisy best-response dynamics

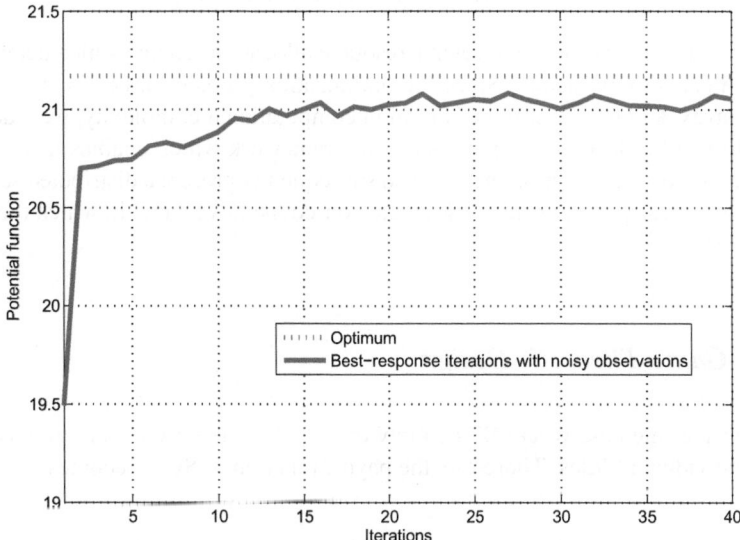

Fig. 4.6 Convergence of the potential function with noisy best-response dynamics

SUs for the CSMA-based MAC design, aiming to reflect the crowding effects in the
activity factor optimization problem.

4.5.1 Problem Formulation

We formulate the activity factor optimization problem to maximize the overall throughput of all SUs which reflects the competition among SUs under constraints of fixed channel requirements for SUs. Different from the optimization problem in (3.8), we set the utility function as $\sum_{k=1}^{N_s} \sum_{i=1}^{N_a} \bar{\beta}_k^i C_k^i (1 - \bar{\beta}_k^i \alpha_i)$ instead of $\sum_{k=1}^{N_s} \sum_{i=1}^{N_a} \beta_k^i C_k^i (1 - \beta_k^i \alpha_i)$. Note that the orthogonal time sharing constraints (i.e., $\sum_{k=1}^{N_s} \beta_k^i \leq 1$, $i = 1, \ldots, N_a$) are not required in this problem since SUs share the idle channels using the adaptive CSMA scheme. Furthermore, since the contention among SUs is reflected in the utility function, there is no need to keep $\sum_{k=1}^{N_s} \beta_k^i \leq 1$, $i = 1, \ldots, N_a$ for contention control among SUs as well. More specifically, the activity optimization problem is given by

$$\max_{\beta} \sum_{k=1}^{N_s} \sum_{i=1}^{N_a} \frac{\beta_k^i C_k^i}{1 + \sum_{\bar{k}=1, \bar{k} \neq k}^{N_s} \beta_{\bar{k}}^i} \left(1 - \frac{\beta_k^i \alpha_i}{1 + \sum_{\bar{k}=1, \bar{k} \neq k}^{N_s} \beta_{\bar{k}}^i} \right) \tag{4.27a}$$

$$\text{subject to } \sum_{i=1}^{N_a} \beta_k^i = R_k, \ k = 1, \ldots, N_s \tag{4.27b}$$

$$0 \leq \beta_k^i \leq 1, \ i = 1, \ldots, N_a, \ k = 1, \ldots, N_s. \tag{4.27c}$$

Based on (4.27), we aim to develop a resource allocation scheme which determines the optimal activity factors. Since the optimization problem in (4.27) is generally non-convex, and hence, suffers from high computational complexity, we study the cognitive MAC design in a game-theoretic framework which enables us to model interactions between competing SUs and subsequently present a distributed scheme. More specifically, we consider a strategic non-cooperative game in which the players are SUs.

4.5.2 Game-Theoretic Design

We consider the case where SUs simply choose their activity factors to maximize their individual utilities. Therefore, the payoff function of SU k becomes

$$u_k' = \sum_{i=1}^{N_a} \frac{\beta_k^i C_k^i}{1 + \sum_{\bar{k}=1, \bar{k} \neq k}^{N_s} \beta_{\bar{k}}^i} \left(1 - \frac{\beta_k^i \alpha_i}{1 + \sum_{\bar{k}=1, \bar{k} \neq k}^{N_s} \beta_{\bar{k}}^i} \right). \tag{4.28}$$

Let $\boldsymbol{\beta}_k = \left[\beta_k^1, \ldots, \beta_k^{N_a}\right]$ be the strategy at SU k and $\boldsymbol{\beta}_{-k}$ be the strategy of all SUs excluding the SU k. Furthermore, the admissible strategies of SU k is defined as

$$\mathscr{D}_k = \left\{ \boldsymbol{\beta}_k : 0 \le \beta_k^i \le 1, \forall i \in \mathscr{N}_a, \sum_{i=1}^{N_a} \beta_k^i = R_k \right\}. \tag{4.29}$$

Then, we can define the non-cooperative game for the MAC design in OSA networks as

$$\mathscr{G}' = \left[\mathscr{N}_s, \{\mathscr{D}_k\}_{k \in \mathscr{N}_s}, \{u_k'\}_{k \in \mathscr{N}_s}\right] \tag{4.30}$$

where $\mathscr{N}_s = \{1, \ldots, N_s\}$ is the set of players of the game (i.e., SUs), $\mathscr{D}_k$ is the activity factor strategy set of the SU k, and u_k' is the corresponding payoff function of the SU k defined on the set of pure-strategy profiles $\mathscr{D} = \mathscr{D}_1 \times \ldots \times \mathscr{D}_{N_s}$.

4.5.2.1 Existence of the Nash Equilibrium

Although the NE concept predicts a stable outcome of a non-cooperative game, such a point does not necessarily exist. Thus, first, we study the existence of NE of the game $\mathscr{G}'$ in the following theorem.

Theorem 4.6 *The game $\mathscr{G}'$ admits at least one pure-strategy NE.*

Proof: This comes directly from Theorem 1 in [10], which presents sufficient conditions for the existence of NE for games with continuous payoff functions. Accordingly, since $\mathscr{D}_k$'s are compact and convex sets, and u_k' is a continuous function and also concave in $\mathscr{D}_k$ (definition of a concave game), the game $\mathscr{G}'$ has at least one pure-strategy NE.∎

4.5.2.2 Uniqueness of the Nash Equilibrium

To investigate the convergence issues, after ensuring the NE existence, it is important to know whether the NE of the game $\mathscr{G}'$ is unique or not. The following theorem investigates the uniqueness of NE of the game $\mathscr{G}'$.

Theorem 4.7 *The pure-strategy NE of the game $\mathscr{G}'$ is unique for non-zero α_i values.*

Proof: Theorem 2 in [10] guarantees NE uniqueness for the concave games if a certain condition, called diagonally strict concavity (DSC), is met. Based on this theorem, we establish the uniqueness for the game $\mathscr{G}'$. A game with strictly concave payoff functions satisfies DSC. Accordingly, since u_k' is a strictly concave function in $\mathscr{D}_k$ under assumption of non-zero α_i's, it can be concluded that $\mathscr{G}'$ has a unique pure-strategy NE.∎

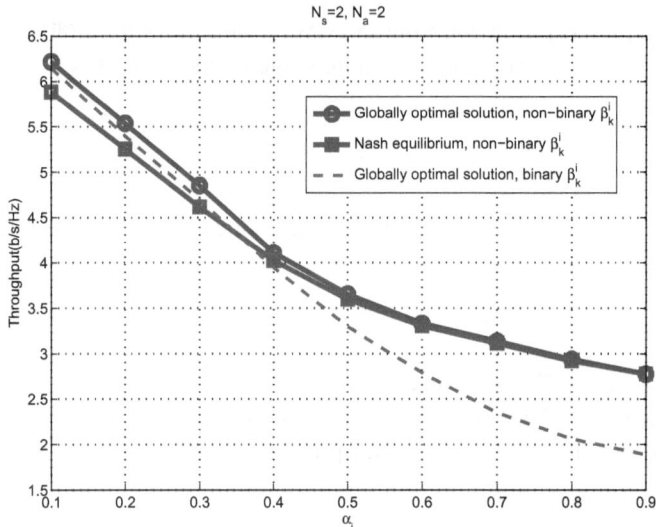

Fig. 4.7 Performance comparison of the NE of game $\mathscr{G}'$ (with non-binary β_k^i) and the globally optimal solution (with non-binary and binary β_k^i) versus the PU return probability α_i for fixed $N_s = 2$, $N_a = 2$ and SNR $= 10$ dB

4.5.2.3 Efficiency of the Nash Equilibrium

The other aspect which is important to characterize is the equilibrium efficiency in a game-theoretic design, since it is the state at which the network will spontaneously operate. In other words, we are interested to study how efficient the NE of game $\mathscr{G}'$ is as compared to the optimal solution of (4.27). To this end, numerical results are provided to compare the overall throughput of SUs which can be obtained from the globally optimal solution of (4.27) to that which can be reached from the NE of game $\mathscr{G}'$.

In this example, we assume independent channels with the same bandwidth $B^i = 1$ and the same α_i. We set the same SNR $= \frac{p_k^i}{n_k^i} = 10$ dB and same channel requirement (i.e., $R_k = 1$) for individual SUs. Figure 4.7 shows that the unique NE of $\mathscr{G}'$ may be inefficient in terms of the total profit for all SUs. However, the throughput decrease is small while the globally optimal solution can only be achieved at the cost of high complexity. Furthermore, it demonstrates the overall throughput of SUs while β_k^i are restricted to binary values. Apparently, the proposed hopping-based MAC strategy (with non-binary β_k^i) improves the performance comparing to the existing OSA MAC approaches (with binary β_k^i). For the higher range of α_i, even the overall throughput obtained in NE of game $\mathscr{G}'$ with non-binary β_k^i is larger than which can be obtained from globally optimal solution of (4.27) with binary β_k^i.

4.5.3 Cooperative Design with Dynamic Pricing

In the non-cooperative game $\mathscr{G}'$, each SU acts self-interestedly, and hence, ignores the cost imposed to the other SUs. Thus, a key challenge is how to modify the payoff function of each SU to deal with NE inefficiency by inducing cooperation, while maintaining the non-cooperative game framework. Pricing has been used as an effective tool to motivate SUs in a non-cooperative game to adopt a more cooperative behavior [11]. By adding a pricing mechanism to the payoff functions, each SU pays a price for using the resources, and hence, voluntarily cooperates with other SUs.

Therefore, we develop a non-cooperative game with pricing (i.e., $\mathscr{G}''$) which is practically the same game as $\mathscr{G}'$ with different payoff functions, as given below,

$$\mathscr{G}'' = \left[\mathcal{N}_s, \{\mathcal{D}_k\}_{k \in \mathcal{N}_s}, \{u_k''\}_{k \in \mathcal{N}_s} \right]. \tag{4.31}$$

In the game $\mathscr{G}''$, considering a linear usage-based pricing scheme, the payoff functions are constructed as

$$u_k'' = \sum_{i=1}^{N_a} \frac{\beta_k^i C_k^i}{1 + \sum_{\bar{k}=1, \bar{k} \neq k}^{N_s} \beta_{\bar{k}}^i} \left(1 - \frac{\beta_k^i \alpha_i}{1 + \sum_{\bar{k}=1, \bar{k} \neq k}^{N_s} \beta_{\bar{k}}^i} \right) - \sum_{i=1}^{N_a} \bar{\mu}_k^i \beta_k^i \tag{4.32}$$

where $\bar{\mu}_k^i \geq 0$ represents the pricing factor of the SU k in channel i.

Remark 4.3 *Since payoff functions with pricing (i.e., u_k'') are also strictly concave for non-zeros α_i values, similar to Theorems 4.6 and 4.7, it is clear that the game $\mathscr{G}''$ owns a unique pure-strategy NE.*

In general, the pricing factors need to be adjusted in such a way that it offers the largest possible enhancement in the overall throughput. In this work, we consider a game with dynamic pricing factors by defining $\bar{\mu}_k^i = \alpha_i \sum_{\bar{k}=1, \bar{k} \neq k}^{N_s} \beta_{\bar{k}}^i$. This implies that each SU pays a penalty for each channel proportional to the crowdedness and the PU return probability of the corresponding channel. As a result, each SU may avoid channels in which the other SUs have already high activity. Thus, this setting can facilitate to resolve contention among SUs in the crowded channels and subsequently improve the NE efficiency in comparison with the game $\mathscr{G}'$ with no pricing.

To investigate the NE efficiency of the game $\mathscr{G}''$, we present numerical results which evaluate the overall throughput of SUs obtained from the NE of game $\mathscr{G}''$. With the same setting as Fig. 4.7, Fig. 4.8 shows that pricing mechanism offers improvement in terms of the total profit for all SUs compared to the game $\mathscr{G}'$ with no pricing. Particularly, for higher α_i, the performance of the NE with pricing is close to the optimal solution.

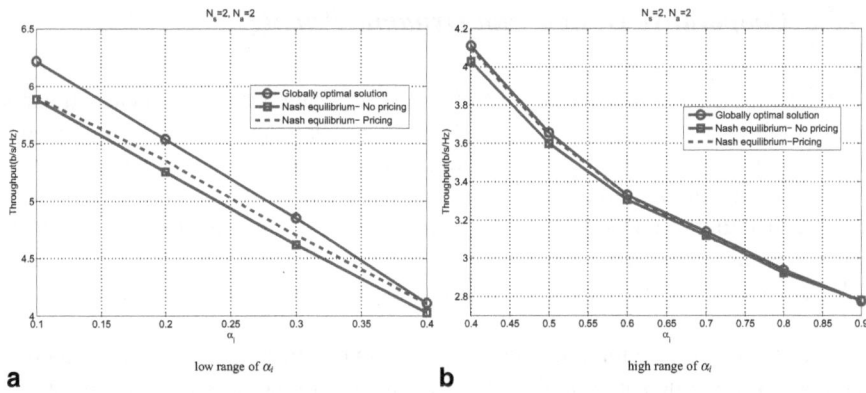

Fig. 4.8 Performance comparison of the NE of game $\mathscr{G}''$ and the globally optimal solution versus the PU return probability α_i for fixed $N_s = 2$, $N_a = 2$ and SNR $= 10$ dB. **a** low range of α_i. **b** high range of α_i

4.5.4 Learning Equilibrium

In this section, we explore how to reach the unique NE of the formulated games as a consequence of a long-run learning process. The key idea is to present an iterative algorithm in which each SU could update its strategy independently while terminating with the unique NE.

One reasonable dynamic learning process is called the continuous best-response dynamics in which each player changes its strategy at a rate proportional to the gradient of its payoff function [10]. Assuming that $\bar{\lambda}_k$ denotes the proportionality constant for the SU k, differential equations for updating activity factors $\boldsymbol{\beta}_k$ become

$$\frac{\partial \boldsymbol{\beta}_k}{\partial t} = \bar{\lambda}_k \nabla_k u'_k, k = 1, \ldots, N_s \tag{4.33}$$

where $\nabla_k u'_k$ denotes the gradient with respect to $\boldsymbol{\beta}_k$ of u'_k. The following theorem ensures the convergence of the continuous best-response dynamics in (4.33) for the games $\mathscr{G}'$ and $\mathscr{G}''$.

Theorem 4.8 *The continuous best-response dynamics converge to the unique pure-strategy NE considering either the game $\mathscr{G}'$ or game $\mathscr{G}''$ from any feasible initial strategy point for non-zero α_i values.*

Proof: This comes directly from Theorem 8 in [10] in which it is shown that for a game satisfying DSC, the system in (4.33) is globally asymptotically stable with respect to the unique NE of the game. Since a game with strictly concave payoff functions satisfies DSC, the continuous best-response dynamics converge to the unique NE for both the strictly concave game $\mathscr{G}'$ and the strictly concave game $\mathscr{G}''$. ∎

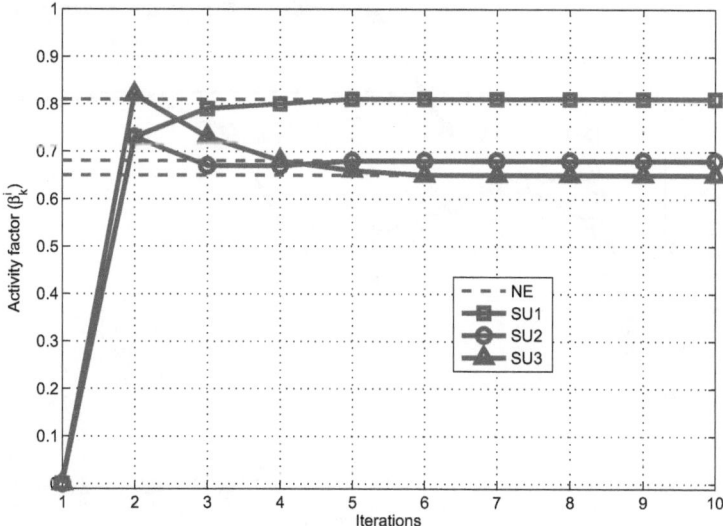

Fig. 4.9 Convergence of the SU activity factors in the proposed iterative algorithm considering the game $\mathscr{G}'$

Consequently, we present iterative game-theoretic algorithms for activity factor selection to reach the unique equilibrium of both game $\mathscr{G}'$ and game $\mathscr{G}''$, based on the best-response dynamics. In particular, in a round robin fashion, SUs iteratively update their activity factors based on the best-response dynamics defined as

$$\boldsymbol{\beta}_k[t] = \arg \max_{\boldsymbol{\beta}'_k \in \mathscr{B}_k} u_k(\boldsymbol{\beta}'_k, \boldsymbol{\beta}_{-k}[t-1]). \tag{4.34}$$

To verify convergence of the iterative algorithm based on (4.34), numerical results are also provided. Figure 4.9 demonstrates the convergence process of the activity factors of three SUs in one of the idle licensed channels considering the game $\mathscr{G}'$. Similarly, Fig. 4.10 shows the convergence for the game $\mathscr{G}''$ with pricing. They confirm that the best-response iterations will converge to the unique NE. As can be observed, the OSA MAC scheme takes merely around 10 iterations to converge to the NE. Note that each iteration corresponds to a complete round-robin update by all the SUs. Furthermore, comparing Fig. 4.10 with Fig. 4.9, it is clear that using the pricing scheme keeps $\sum_{k=1}^{N_s} \beta_k^i$ smaller that means less contention among SUs in the channel i. By decreasing contention among SUs in a specific channel and distributing SUs' activity over different channels, the pricing scheme improves the total throughput.

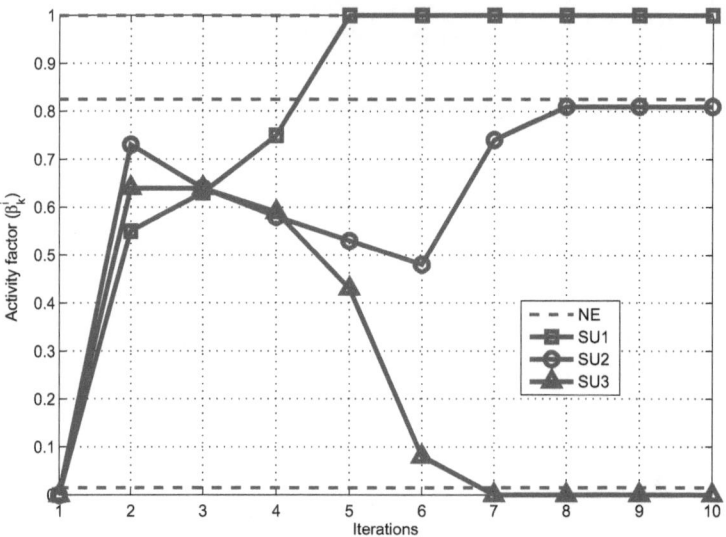

Fig. 4.10 Convergence of the SU activity factors in the proposed iterative algorithm considering the game $\mathscr{G}''$

4.6 Concluding Remarks

In this chapter, we have developed a distributed game-theoretic CSMA-based MAC scheme in which activity factors of SUs over multiple idle channels are adaptively adjusted. Via potential game framework, it is established that the formulated game admits at least one pure-strategy NE. In consideration of coupling constraints among SUs, sufficient conditions are presented to ensure the feasibility of the pure-strategy NE. In addition, it is proved that the globally optimal solution is the Pareto-optimal NE.

Furthermore, we have investigated the convergence properties of the best-response dynamics and log-linear dynamics of the formulated game. Assuming that the perfect knowledge of moves previously made by all SUs is available for each SU, we have proved that best-response iterations converge to a pure-strategy NE which is not essentially the global solution. However, the log-linear process enables equilibrium refinement and convergence to the most socially desirable solution. Subsequently, in a game with noisy observations, we have shown that best-response iterations also converge with probability of 1 to a neighborhood of the global optimum.

Moreover, we have presented an alternative design objective for SUs which reflects competition among SUs and formulated the problem in the game-theoretic framework. Via concave game framework, we have established that the formulated game admits a unique pure-strategy NE which is not necessarily efficient. With the aid of a dynamic pricing mechanism, we have improved the NE efficiency by inducing cooperation in the non-cooperative game. Furthermore, we have proved that the iterative algorithms based on the best-response dynamics converge to the unique pure-strategy NE.

References

1. C. Alos-Ferrer and N. Netzer, "The logit-response dynamics," *Games and Economic Behavior*, vol. 68, no. 2, pp. 413–427, Mar. 2010.
2. J. Marden and J. Shamma, "Revisiting log-linear learning: Asynchrony, completeness and a payoff-based implementation," *Games and Economic Behavior*, vol. 75, no. 2, pp. 788–808, Jul. 2010.
3. D. Monderer and L. Shapley, "Potential games," *Games and Economic Behavior*, vol. 14, no. 1, pp. 124–143, 1996.
4. A. B. MacKenzie, L. Dasilva, and W. Tranter, *Game Theory for Wireless Engineers*. Morgan and Claypool Publishers, 2006.
5. G. Scutari, S. Barbarossa, and D. P. Palomar, "Potential games: A framework for vector power control problems with coupled constraints," in *Proc. IEEE Int. Conf. on Acoust., Speech and Signal Process. (ICASSP)*, Toulouse, France, May 2006.
6. D. Fudenberg and D. K. Levine, *The Theory of Learning in Games*. MIT press, 1998, vol. 2.
7. D. Okada and O. Tercieux, "Log-linear dynamics and local potential," *Journal of Economic Theory*, vol. 147, no. 3, pp. 1140–1164, May 2012.
8. S. Lasaulce and H. Tembine, *Game Theory and Learning for Wireless Networks: Fundamentals and Applications*. Academic Press, 2011.
9. R. Menon, A. B. MacKenzie, J. Hicks, R. M. Buehrer, and J. H. Reed, "A game-theoretic framework for interference avoidance," *IEEE Trans. Commun.*, vol. 57, no. 4, pp. 1087–1098, Apr. 2009.
10. J. B. Rosen, "Existence and uniqueness of equilibrium points for concave n-person games," *Econometrica: Journal of the Econometric Society*, vol. 33, no. 3, pp. 520–534, Jul. 1965.
11. C. Saraydar, N. B. Mandayam, and D. J. Goodman, "Efficient power control via pricing in wireless data networks," *IEEE Trans. Commun.*, vol. 50, no. 2, pp. 291–303, Feb. 2002.

Chapter 5
Adaptive Carrier Sensing-Based MAC Designs: Throughput Analysis

5.1 Introduction

In Chaps. 3 and 4, we have presented an adaptive CSMA scheme, aiming to present fully-distributed MAC algorithms (including non-game-theoretic and game-theoretic approaches) for SUs in OSA networks. In the proposed adaptive CSMA, each SU that has a new packet for transmission enters a competition (i.e., backoff mechanism) to access an idle channel with a certain probability (called activity factor). Activity factors of different SUs need to be optimized based on channel qualities, PU return probabilities and sharing incentives. By assigning adaptive access probabilities to different SUs, the proposed adaptive CSMA prioritizes SUs that will gain most from using a channel, and hence, improves channel utilization compared to a simple random access scheme. Although the proposed random access scheme (adaptive CSMA) aims to minimize the collision probability among SUs, the contention among different STAs is inherent in any random access scheme. Thus, in this chapter, we develop an analytical model to compute the system throughput and evaluate the performance of the adaptive CSMA in the presence of inevitable collisions.

More specifically, we analyze the collision probability among competing STAs (i.e., SUs) in a network and study the saturation throughput of the proposed adaptive CSMA in a single idle channel. In [1], the saturation throughput of a CSMA scheme is defined as the fraction of opportunities which are used successfully to transmit data, assuming that STAs always have data to transmit. Comparing to the conventional CSMA, we show that the adaptive CSMA significantly decreases the collision probability and increases the saturation throughput, specifically in networks with larger number of STAs. Furthermore, we investigate the effects of the network configuration-based adaptations on the saturation throughput.

To study the saturation throughput of the adaptive CSMA, we design a MAC layer which can be backward compatible with distributed coordination function (DCF), IEEE 802.11 CSMA-based MAC mechanism. However, the random backoff mechanism is slightly modified to mitigate the contention among STAs and improve

© The Author(s) 2014   83
M. Derakhshani, T. Le-Ngoc, *Cognitive MAC Designs for OSA Networks*,
SpringerBriefs in Electrical and Computer Engineering, DOI 10.1007/978-3-319-12649-4_5

the throughput performance. Similar to DCF, we consider the binary exponential backoff rules and the collision avoidance operations to manage retransmission of collided packets.

In the IEEE 802.11 MAC enhancement studies, there are two different directions including collision avoidance overhead reduction (e.g., [2–7]) or/and collision probability decrease (e.g., [8–12]).

On one hand, to decrease the collision avoidance overhead, IEEE 802.11e amendment offers frame bursting in which the STA that obtains transmission opportunity (after winning in the backoff competition) can send a burst of back-to-back packets based on its channel quality [3, 4]. Furthermore, MAC frame aggregation is introduced in IEEE 802.11n to decrease the frequency of PHY and MAC overheads during transmission of multiple packets [5]. In the frame aggregation, by concatenating or packing multiple packets together, overheads can be added over a group of packets rather than over separate ones [5].

On the other hand, to reduce the collision probability, there are works that focuse on optimizing the backoff algorithm of CSMA to reduce the collision probability, and hence, to maximize the overall throughput. For instance, in [8,9], the contention window size is optimized depending on the network and load configuration (e.g., number of active STAs and packet length) to control contention among STAs. In [10], a collision-minimizing CSMA is proposed in which the probability distribution using which STAs randomly select their backoff time is not necessarily uniform; rather, it is carefully chosen to minimize the collision among the competing STAs.

Furthermore, [11], which is the closest in spirit with our proposed adaptive CSMA, has proposed an opportunistic CSMA scheme for a WLAN to improve throughput by exploiting the multi-user diversity gain. The proposed opportunistic CSMA prioritizes the STAs with high-SNRs by granting earlier access to those STAs. Since the priority is given by earlier access, there is a conflict between STAs with high-SNRs and STAs who are newly arrived with smaller contention window sizes. To address this issue, it is assumed that all STAs share the same contention window size. By enforcing same contention window size among different STAs, backoff operation needs to be performed centrally at the AP and the resulting backoff window size needs to be broadcasted to all STAs. However, in our work, prioritizing STAs with a good channel quality is based on granting higher access probability, instead of earlier access as in [11]. Consequently, the proposed adaptive CSMA can support fully distributed and asynchronous operation with the exponential backoff mechanism on the STA-side.

The remainder of this chapter is organized as follows. In Sect. 5.2, the system model and problem formulation under consideration are provided. Next, Sect. 5.3 analyzes the saturation throughput of the proposed adaptive CSMA. Furthermore, in Sect. 5.4, numerical results are provided to validate the throughput analysis and illustrate throughput improvement compared to the conventional CSMA. Finally, Sect. 5.5 presents the concluding remarks.

5.2 System Model and Problem Formulation

Different from DCF, the proposed adaptive CSMA attempts to manage the contention among STAs prior to entering the backoff mechanism. To enable such contention management, it assigns a certain probability (called activity factor) to each STA with a new packet for transmission and allows it to enter the backoff competition to access a specific idle channel (e.g., i) based on its own activity factor, $\beta_k^i (0 \le \beta_k^i \le 1)$. Adaptive activity factors enable prioritizing STAs who gain most from using a channel, and hence, improving the channel utilization in comparison with a simple random access scheme.

More specifically, in the adaptive CSMA, the STA k–which already sensed the channel i idle for a DIFS and is ready to transmit a packet–performs the following steps:

1. Generate a Bernoulli random variable x_k^i with the success probability β_k^i. If $x_k^i = 0$, the STA k will not transmit and defer its transmission for a period of time called long inter-frame space (LIFS). If $x_k^i = 1$, the STA k will proceed to the next step.
2. Generate a backoff time according to a uniform distribution in the interval $(0, W - 1)$.
3. After expiry of the backoff time, sense the channel i, if it is idle, continue to transmit using the basic access or the RTS/CTS access mechanism rules.

Note that LIFS is a newly-defined inter-frame space parameter in the adaptive CSMA scheme. We assume that LIFS is properly designed such that the STA with $x_k^i = 0$ could skip the current competition and retry after the on-going packet transmission process. For instance, LIFS can be set equal to W for each STA, to ensure that the corresponding STA with $x_k^i = 0$ would not capture the channel.

According to the activity factor optimization problem in (3.8), it is not possible to derive the closed-form expressions of the optimal activity factors based on the channel qualities, the PU return probabilities and the number of SUs. Thus, to study the throughput performance of the adaptive CSMA, we consider a specific case in which the adaptive CSMA attempts to take advantage of the channel diversity among different STAs and give higher chance of access to the STAs with better channel qualities. To this end, inspired by (3.13), the activity factor of each STA in each channel is defined as an increasing function of its channel transmission capacity as follows,

$$\beta_k^i = \left[1 - \frac{\tilde{\mu}_k^i}{C_k^i} \right]_0^1 \tag{5.1}$$

where $[x]_a^b = \min(b, \max(a, x))$, C_k^i represents the channel transmission capacity for the STA k in the channel i, and $\tilde{\mu}_k^i$ denotes the threshold for the STA k in the channel i which is a positive scalar. Note that the derived analytical results on the saturation throughput–presented in the next section–are generally developed as functions of

β_k^i, and are not dependent on the specific definitions of β_k^i. The reason for the specific definition of β_k^i in (5.1) is to be used in the numerical results to illustrate the throughput performance.

Similar to Chaps. 3 and 4, the channel transmission capacity of the STA k in the channel i is represented by $C_k^i = B^i \log(1 + P_k^i g_{k,k}^i / n_k^i)$ where B^i denotes the bandwidth in the channel i, $\frac{P_k^i}{n_k^i}$ is the signal-to-noise ratio (SNR) for the STA k in the channel i, and $g_{k,k}^i$ is the channel power gain for the STA k in the channel i. In this work, a block-fading model is assumed in which $g_{k,k}^i$ remains unchanged within the coherence time but independently varies of the previous channel realization. To make sure that $g_{k,k}^i$ does not change during a packet transmission, the coherence time is assumed sufficiently long in this work. Furthermore, the channel power gains $g_{k,k}^i$ are considered independent for different STAs.

To support the distributed operation by each STA and allow the fair access among different STAs, we assume that each STA updates its activity factor in each channel relative to its own average channel transmission capacity. Then, we define the threshold for the STA k in the channel i as

$$\tilde{\mu}_k^i = \rho_k^i \mathrm{E_g}[C_k^i] \tag{5.2}$$

where ρ_k^i is a positive scalar and $\mathrm{E_g}[.]$ denotes the expectation with respect to the channel power gain distribution. In practice, this expected value (i.e., $\mathrm{E_g}[C_k^i]$) can be empirically estimated by each STA. By adjusting the threshold based on (5.2), each STA gets a higher chance of transmission if its channel transmission capacity (i.e., C_k^i) is relatively higher than its own average (i.e., $\mathrm{E_g}[C_k^i]$). In other words, each STA is compared with itself, and hence obtains the fair access over a long time.

In particular, to enable a long-term fairness for different STAs with different channel conditions (i.e., different SNR $= \frac{P_k^i}{n_k^i}$ and/or different probability distribution for $g_{k,k}^i$), we select ρ_k^i for each STA in each channel such that the same average access probability (i.e., $\mathrm{E_g}[\beta_k^i] = \beta$) can be achieved for all STAs. More specifically,

$$\mathrm{E_g}[\beta_k^i] = \mathrm{E_g}\left\{\left[1 - \frac{\rho_k \mathrm{E_g}[C_k^i]}{C_k^i}\right]_0^1\right\} = \beta. \tag{5.3}$$

According to (5.1), (5.2) and (5.3), it is clear that each STA can update its activity factor in each channel based on the locally available information and does not need to know the knowledge on other STAs. Thus, the proposed adaptive CSMA can be implemented in a fully-distributed manner.

Furthermore, aiming to enhance the efficiency of the proposed adaptive CSMA, ρ_k^i can be tuned dynamically depending on the network-configuration parameters. Some network parameters such as the number of STAs in the network have a significant impact on the system throughput. Assuming that STAs have the perfect knowledge of the number of STAs in the network, ρ_k^i can be defined as an increasing function of N_s to control $\sum_{k=1}^{N_s} \beta_k^i$ (i.e., the sum of all activity factors in the channel i) as N_s increases. In other words, by increasing the thresholds (i.e., ρ_k^i) in the larger networks, the adaptive CSMA attempts to reduce the activity factors of STAs. Such dynamic threshold setting helps to control the contention among STAs and decreases the collision probability. For instance, assuming that STAs have same SNR, ρ_k^i can be chosen as $1 - \frac{1}{N_s}$. In the next section, we investigate the effects of choosing different ρ_k^i on the throughput by the numerical results.

5.3 Throughput Analysis

In this section, we study the saturation throughput of the proposed adaptive CSMA in a single idle channel (e.g., i), assuming a constant packet size L_P. In [1], the saturation throughput for CSMA-based DCF is defined and calculated as the number of successfully delivered information bits per second, while the transmission queue of each STA is assumed to be always nonempty. Accordingly, considering that the activity factors of different STAs are channel dependent, we define the saturation throughput of the adaptive CSMA as

$$T_{\text{saturation}} = E_g \left[\frac{P_s L_P}{(1 - P_{\text{tr}})T_b + \sum_{k=1}^{N_s} P_{s,k} T_{s,k} + (P_{\text{tr}} - P_s)T_c} \right] \tag{5.4}$$

where T_b is the duration of a backoff time-slot, $P_s = \sum_{k=1}^{N_s} P_{s,k}$ is the probability that a successful transmission happens in a generic (i.e., randomly chosen) backoff time-slot, $P_{s,k}$ is the probability of a successful transmission by the STA k in a generic backoff time-slot, P_{tr} is the probability that at least one transmission, either successful or not, happens in a generic backoff time-slot, $T_{s,k}$ is the average time that the channel is sensed busy because of a successful transmission by the STA k, and T_c is the average time that the channel is sensed busy by each STA because of a collision.

In the saturation throughput expression (5.4), the numerator represents the average number of information bits that successfully transmitted in a backoff time-slot, i.e., $P_s L_P$. In the denominator, the average length of a backoff time-slot is presented. In particular, the back-off time-slot is empty with probability $(1 - P_{\text{tr}})$, it contains a successful transmission with probability P_s, and it contains a collision with probability $P_c = P_{\text{tr}} - P_s$ [1]. To analyze the saturation throughput in (5.4), first, we need to derive the probability (called access probability) that a single STA transmits a packet in a generic backoff time-slot. Then, by defining P_s and P_{tr}, we can express the throughput as a function of access probability.

For the *conventional CSMA* scheme, in [1], the access probability P_a of a STA has been studied by analyzing the behavior of a single STA with a Markov model. The proposed Markov chain models the binary exponential backoff rules and the collision avoidance operations of a single STA. Assuming that each transmitted packet collides and fails with a constant and independent probability (P_f), the access probability of a STA is computed as a function of P_f, the contention window W and the maximum backoff stage m as follows

$$P_a = \frac{2(1-2P_f)}{(1-2P_f)(W+1)+P_f W(1-(2P_f)^m)}. \tag{5.5}$$

On the other hand, the probability that a transmitted packet in a generic backoff time-slot encounters a collision is equal to the probability that at least one of the remaining STAs transmits. Thus,

$$P_f = 1-(1-P_a)^{N_s-1} \tag{5.6}$$

where N_s denotes the number of competing STAs [1]. By solving the nonlinear system of (5.5) and (5.6), P_a can be obtained as a function of N_s, W and m. In [1], it is proved that the nonlinear system of P_a and P_f has a unique solution.

In the proposed *adaptive CSMA* scheme, the access probability of the STA k is different and can be calculated as $\beta_k^i P_a$ in consideration of the adaptive access scheme based on the activity factors. Then, since there are N_s STAs in the network attempting to transmit on the channel, each with different access probability $\beta_k^i P_a$, the probability that there is at least one transmission in a generic backoff time-slot (i.e., P_{tr}) is

$$P_{tr} = 1-\prod_{k=1}^{N_s}(1-\beta_k^i P_a). \tag{5.7}$$

A transmitted packet will be received successfully, if exactly one STA transmits on the channel. Thus, the probability of successful transmission (i.e., P_s) becomes

$$P_s = \sum_{k=1}^{N_s} P_{s,k} = \sum_{k=1}^{N_s} \beta_k^i P_a \prod_{k'=1,k'\neq k}^{N_s} (1-\beta_{k'}^i P_a). \tag{5.8}$$

Based on (5.4), to specifically compute the saturation throughput, it is required to specify the values of $T_{s,k}$ and T_c. Consider that H represents the size of PHY and MAC headers and δ denotes the propagation delay. In the basic access mechanism, we have

$$T_{s,k}^{bas} = H + \frac{L_P}{C_k^i} + SIFS + \delta + ACK + DIFS + \delta, \tag{5.9}$$

$$T_c^{bas} = H + L_P^* + DIFS + \delta \tag{5.10}$$

where $\frac{L_P}{C_k^i}$ represents the transmission duration of the STA k and L_P^* represents the average time of the longest packet transmission involved in a collision. Consider a case that different STAs have the same SNR and same probability distribution for $g_{k,k}^i$. In this case, assuming that the collision probability of three or more packets is

Packet payload (L_P)	2400 bytes
Backoff time-slot (T_b)	9 μs
PHY and MAC header (H)	65 μs
Propagation delay (δ)	1μs
SIFS	10 μs
DIFS	28 μs
ACK	40 μs
RST	48 μs
CTS	40 μs
CW_{min}	16
CW_{max}	1024

Table 5.1 IEEE 802.11 MAC parameters used in the numerical results

negligible, L_P^* can be approximated as

$$L_P^* \simeq E_g \left[\max \left\{ \frac{L_P}{C_k^i}, \frac{L_P}{C_{k'}^i} \right\} \right]. \tag{5.11}$$

On the other hand, by using the RTS/CTS access mechanism, collision can happen only during the RTS frames, and hence, $T_{s,k}$ and T_c can be represented as

$$T_{s,k}^{rts} = RTS + SIFS + \delta + CTS + SIFS + \delta + H + \frac{L_P}{C_k^i}$$
$$+ SIFS + \delta + ACK + DIFS + \delta, \tag{5.12}$$

$$T_c^{rts} = RTS + DIFS + \delta. \tag{5.13}$$

Consequently, based on (5.7), (5.8), (5.9), (5.10), (5.12) and (5.13), the saturation throughput can be obtained using (5.4).

5.4 Numerical Results

In this section, we discuss a numerical example on the throughput of the adaptive CSMA scheme based on the analytical and the simulation results. Such numerical example helps us to validate the throughput analysis of the adaptive CSMA, in comparison with the simulation results. Furthermore, it illustrates the throughput improvement offered by the adaptive CSMA relative to the conventional CSMA scheme. For the simulation results, we used a simplified MAC layer simulator which is implemented in Matlab.

The set of parameters used in the numerical results, by both analysis and simulation, are summarized in Table 5.1 based on the IEEE 802.11 g MAC specifications [13]. The channel bandwidth is assumed equal to $B^i = 20$ MHz. Unless

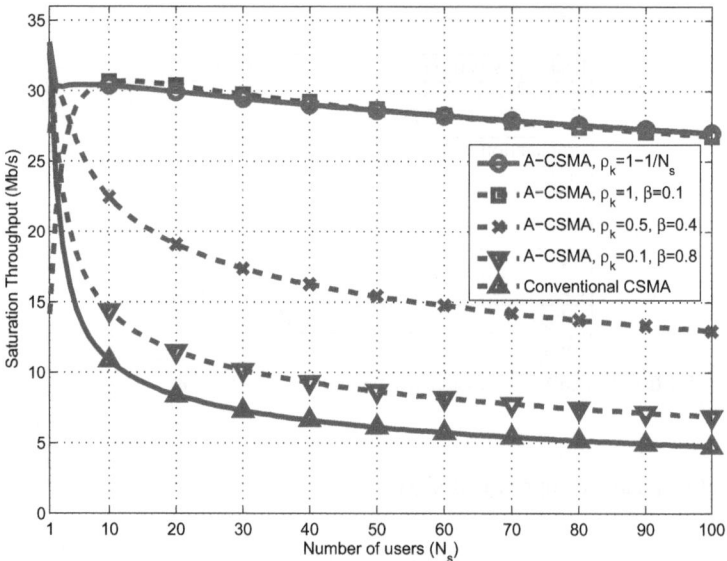

Fig. 5.1 Saturation throughput versus number of STAs using the basic access mechanism

otherwise specified, we consider the same average signal-to-noise ratio (i.e., SNR $= \frac{P_k^i}{n_k^i} = 10$ dB) for all STAs. Furthermore, in all the presented results, the channel power gains $g_{k,k}^i$ are randomly generated according to the Rayleigh distribution assuming $E_g[g_{k,k}^i] = 1$.

Using the basic access mechanism, Fig. 5.1 demonstrates the saturation throughput (i.e., $T_{\text{saturation}}$) of the adaptive CSMA scheme versus the number of STAs in the network (i.e., N_s). In this figure, the throughput performance of the adaptive CSMA is shown considering different threshold settings (i.e., ρ_k^i). Comparing to the conventional CSMA scheme, it is shown that the adaptive CSMA offers a significant throughput improvement for networks with more than 5 STAs. As evident from Fig. 5.1, the adaptive CSMA with the dynamic threshold $\rho_k^i = 1 - \frac{1}{N_s}$ outperforms the conventional CSMA for any number of STAs.

Using the RTS/CTS access mechanism, Fig. 5.2 illustrates the saturation throughput of the adaptive CSMA scheme versus the number of STAs in the network for different threshold settings. As can be observed, the adaptive CSMA with the dynamic threshold $\rho_k^i = 1 - \frac{1}{N_s}$ achieves the multi-user diversity gain using the RTS/CTS access mechanism. In other words, the adaptive CSMA improves the throughput as N_s increases, while the throughput of the conventional CSMA decreases with the number of STAs due to the more frequent collisions. Comparing to Fig. 5.1, it is clear that the RTC/CTS technique improves the throughput performance relative to the basic access scheme. This can be explained by the fact that the RTC/CTS technique effectively reduces the collision time (i.e., T_c). Figures. 5.1 and 5.2 confirm

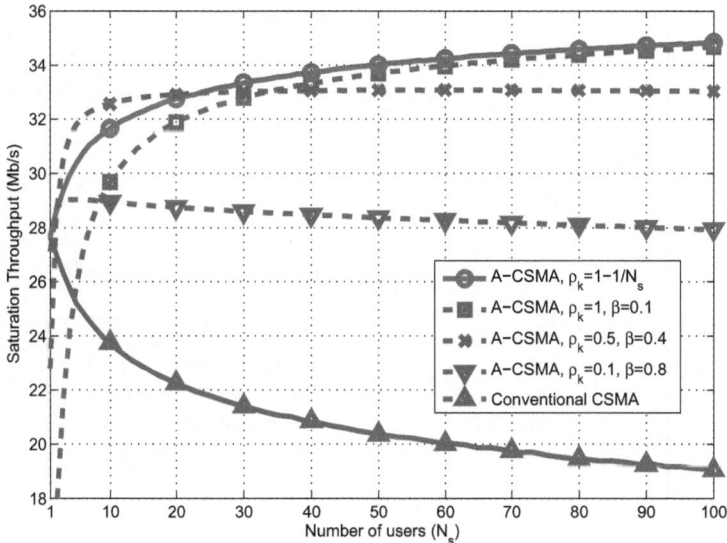

Fig. 5.2 Saturation throughput versus number of STAs using the RTS/CTS access mechanism

the precision of the analytical results on the saturation throughput (lines) as they closely match the simulation results (the symbols), for both the basic access and the RTS/CTS access mechanisms.

Figure 5.3 shows the collision probability (i.e., $P_c = P_{tr} - P_s$) of the adaptive CSMA scheme versus the number of STAs for different threshold settings. It is clear that the adaptive CSMA significantly decreases the collision probability. In agreement with the throughput results in Fig. 5.2, it also demonstrates that the adaptive CSMA with $\rho_k^i = 1 - \frac{1}{N_s}$ provides the largest decrease as compared to the conventional CSMA, except for $\rho_k^i = 1$. Despite the smaller collision probability for $\rho_k^i = 1$, the adaptive CSMA with $\rho_k^i = 1 - \frac{1}{N_s}$ provides a higher throughput. According to Fig. 5.4, this can be explained with the larger probability of successful transmission for the adaptive CSMA with $\rho_k^i = 1 - \frac{1}{N_s}$ comparing to the adaptive CSMA with $\rho_k^i = 1$, specifically for smaller N_s.

To investigate the effects of the packet length on the throughput performance, Fig. 5.5 illustrates the saturation throughput of the adaptive CSMA scheme with $\rho_k^i = 1 - \frac{1}{N_s}$ versus the packet length for $N_s = 20$. Apparently, the saturation throughput improves when the packet length increases. This happens because the MAC overhead is constant, while the packet length is increasing. Furthermore, it is shown that the throughput improvement offered by the adaptive CSMA comparing to the conventional CSMA is an increasing function of the packet length.

Figures 5.6 and 5.7 explore the dependency of the saturation throughput of the adaptive CSMA scheme on the minimum contention window size CW_{min}, using the basic access and the RTS/CTS access mechanisms. In both figures, we assume the

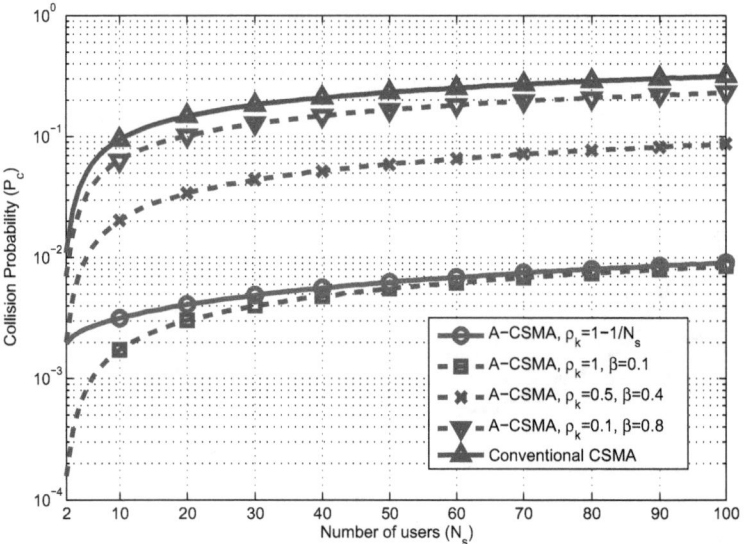

Fig. 5.3 Collision probability versus number of STAs using the RTS/CTS access mechanism

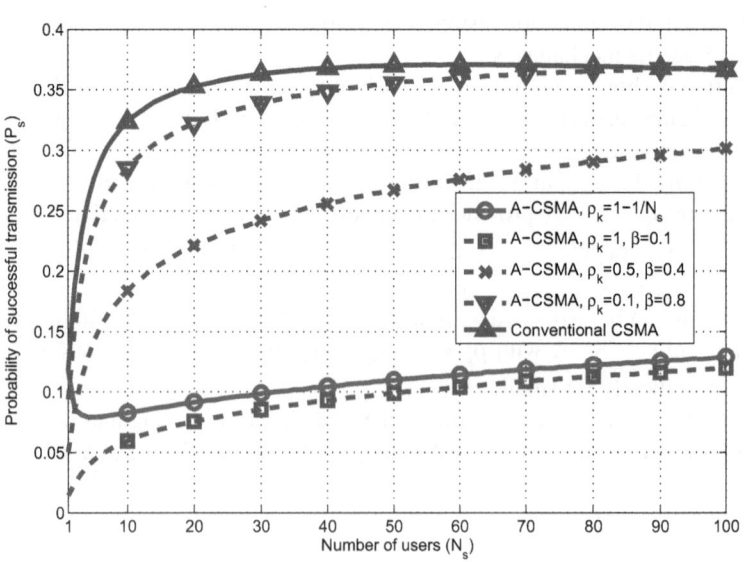

Fig. 5.4 Probability of successful transmission versus number of STAs using the RTS/CTS access mechanism

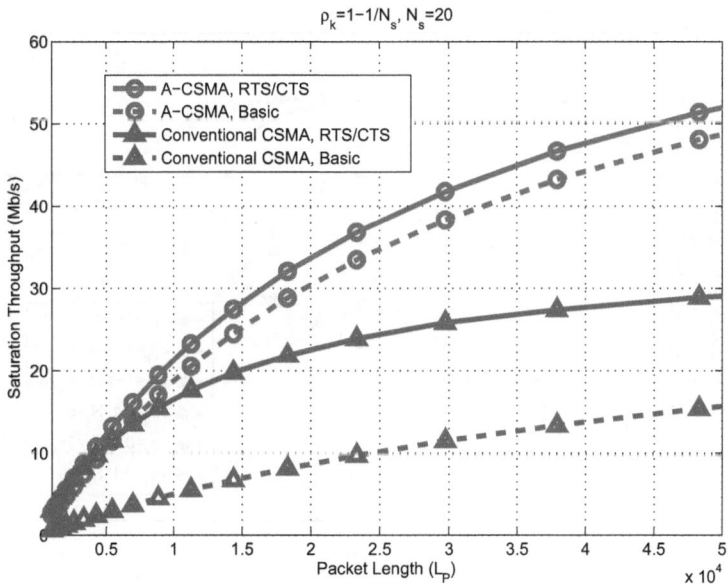

Fig. 5.5 Saturation throughput versus packet length with $N_s = 20$

maximum backoff stage equal to 6 (i.e., $m = 6$). In addition, in each figure, two different threshold settings ($\rho_k^i = 0.1$ and $\rho_k^i = 1 - \frac{1}{N_s}$) are investigated for three different network sizes ($N_s = 5$, $N_s = 25$ and $N_s = 50$).

Using the basic access mechanism, Fig. 5.6 shows that the throughput is highly dependent on the minimum contention window size. To achieve the maximum throughput, it is clear that $CW_{\min}$ needs to be designed as a function of the number of STAs in the network. For instance, the optimal value of $CW_{\min}$ is around 16 for a network with 5 STAs, while $CW_{\min} = 128$ gives a better throughput performance when $N_s = 50$. Furthermore, for a certain number of STAs, it is shown that using a fixed ρ_k reduces the maximum achievable throughput comparing to the dynamic threshold setting $\rho_k = 1 - \frac{1}{N_s}$.

Figure 5.7 shows the behavior of the saturation throughput of the adaptive CSMA scheme with the RTS/CTS access mechanism for different values of $CW_{\min}$. Apparently, in this case, the saturation throughput is less sensitive to the minimum contention window size for the lower range of $CW_{\min}$. For instance, using $\rho_k^i = 0.1$, the saturation throughput is almost independent of the minimum contention window size for $CW_{\min} < 64$. Comparing two different threshold settings, it is shown that the maximum achievable throughput is higher for $\rho_k^i = 1 - \frac{1}{N_s}$.

To confirm that the proposed adaptive CSMA gives fair access to different STAs, Table 5.2 shows the distribution of successfully transmitted packets among 10 STAs, each with a different SNR between 0 and 25 dB. It is shown that the successfully transmitted packets are fairly distributed among STAs.

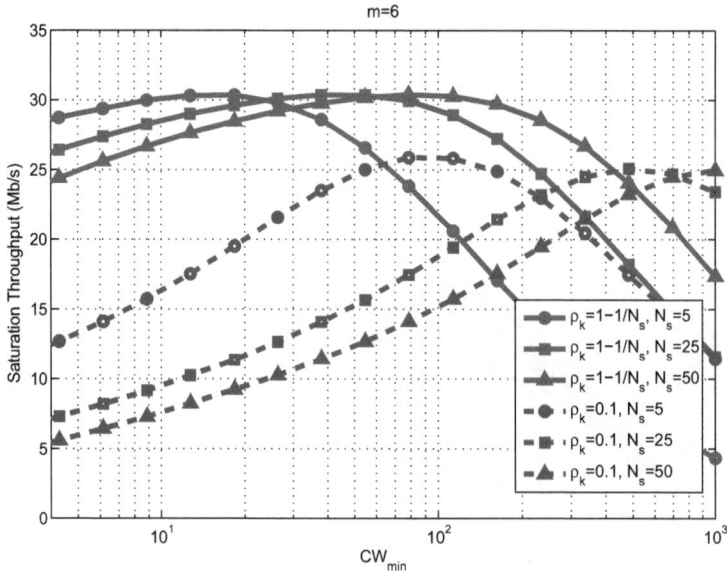

Fig. 5.6 Saturation throughput versus minimum contention window size using the basic access mechanism

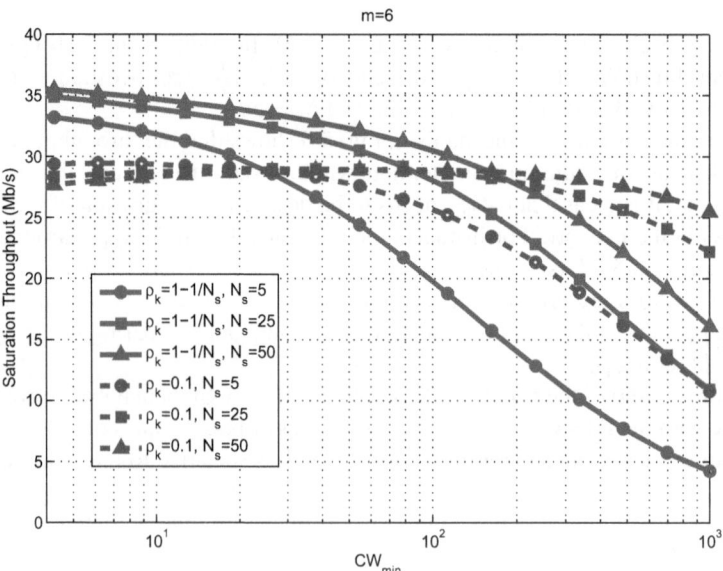

Fig. 5.7 Saturation throughput versus minimum contention window size using the RTS/CTS access mechanism

STA index	1	2	3	4	5	6	7	8	9	10
SNR (dB)	0	2.5	5	7.5	10	12.5	15	17.5	20	25
Percentage (%)	9.96	9.95	9.97	9.93	9.97	9.99	9.96	9.97	9.96	10

Table 5.2 Percentage of successfully transmitted packets for different STAs with different SNRs, assuming $\beta = 0.8$

5.5 Concluding Remarks

In this chapter, we have presented a study on the throughput performance of the proposed adaptive CSMA in a single idle channel. The effects of network configuration parameters (e.g., the number of STAs in the network, the minimum contention window, and the packet length) have been investigated on the saturation throughput. In the adaptive CSMA, STAs go through a refinement process based on their adaptive activity factors, before participating in the backoff competition. As a result, we have shown that the collision probability is decreased due to the smaller number of competitors. Furthermore, the saturation throughput is improved since STAs with better channel qualities are given higher chance to stay in the competition. Numerical results confirm the performance gains, i.e., high throughput as well as long-term fairness, offered by the proposed adaptive CSMA-based access strategy in comparison with the conventional CSMA scheme.

References

1. G. Bianchi, "Performance analysis of the IEEE 802.11 distributed coordination function," *IEEE J. Sel. Areas Commun.*, vol. 18, no. 3, pp. 535–547, Mar. 2000.
2. E. Charfi, L. Chaari, and L. Kamoun, "PHY/MAC enhancements and QoS mechanisms for very high throughput WLANs: A survey," *IEEE Commun. Surveys Tuts.*, vol. 15, no. 4, pp. 1714–1735, Fourth Quarter 2013.
3. I. Tinnirello and S. Choi, "Efficiency analysis of burst transmissions with block ACK in contention-based 802.11e WLANs," in *Proc. IEEE Intl. Conf. Commun. (ICC)*, Seoul, Korea, May 2005.
4. P. K. Hazra and A. De, "Performance analysis of IEEE 802.11e EDCA with QoS enhancements through TXOP based frame-concatenation and block-acknowledgement," *Intl. J. Adv. Tech.*, vol. 2, no. 4, pp. 542–560, 2011.
5. D. Skordoulis, Q. Ni, H. Chen, A. Stephens, C. Liu, and A. Jamalipour, "IEEE 802.11n MAC frame aggregation mechanisms for next-generation high-throughput WLANs," *IEEE Wireless Commun. Mag.*, vol. 15, no. 1, pp. 40–47, Feb. 2008.
6. Y. Lin and V. W. Wong, "WSN01-1: Frame aggregation and optimal frame size adaptation for IEEE 802.11 n WLANs," in *Proc. IEEE Global Commun. Conf. (GLOBECOM)*, San Francisco, CA, USA, Nov. 2006.
7. Y. Kim, S. Choi, K. Jang, and H. Hwang, "Throughput enhancement of IEEE 802.11 WLAN via frame aggregation," in *Proc. IEEE Veh. Tech. Conf. (VTC)*, Los Angeles, CA, USA, Sep. 2004.
8. F. Cali, M. Conti, and E. Gregori, "Dynamic tuning of the IEEE 802.11 protocol to achieve a theoretical throughput limit," *IEEE/ACM Trans. Netw.*, vol. 8, no. 6, pp. 785–799, Dec. 2000.

9. F. Cali, M. Conti, and E. Gregori, "IEEE 802.11 protocol: Design and performance evaluation of an adaptive backoff mechanism," *IEEE J. Sel. Areas Commun.*, vol. 18, no. 9, pp. 1774–1786, Sep. 2000.

10. Y. C. Tay, K. Jamieson, and H. Balakrishnan, "Collision-minimizing CSMA and its applications to wireless sensor networks," *IEEE J. Sel. Areas Commun.*, vol. 22, no. 6, pp. 1048–1057, Aug. 2004.

11. C. S. Hwang and J. M. Cioffi, "Opportunistic CSMA/CA for achieving multi-user diversity in wireless LAN," *IEEE Trans. Wireless Commun.*, vol. 8, no. 6, pp. 2972–2982, Jun. 2009.

12. H. Kwon, S. Kim, and B. Lee, "Opportunistic multi-channel CSMA protocol for OFDMA systems," *IEEE Trans. Wireless Commun.*, vol. 9, no. 5, pp. 1552–1557, May 2010.

13. IEEE 802.11g WG, "Wireless LAN medium access control (MAC) and physical layer (PHY) specifications: Further higher data rate extension in the 2.4 GHz band," Jan. 2003.